HOW TO REPLACE & INSTALL ROOFS & SIDINGS

Created and designed
by the editorial staff
of ORTHO books

Project Editor	Sally W. Smith
Writer	T. Jeff Williams
Designer	Jacqueline Jones
Illustrator	Ron Hildebrand
Photographer	Laurie A. Black

Ortho Books

Publisher
Robert L. Iacopi

Editorial Director
Min S. Yee

Managing Editors
Anne Coolman
Michael D. Smith

System Manager
Mark Zielinski

Senior Editor
Sally W. Smith

Editors
Jim Beley
Diane Snow
Deni Stein

System Assistant
William F. Yusavage

Production Manager
Laurie Sheldon

Photographers
Laurie A. Black
Michael D. McKinley

Photo Editors
Anne Dickson-Pederson
Pam Peirce

Production Editor
Alice E. Mace

Production Assistant
Darcie S. Furlan

National Sales Manager
Garry P. Wellman

Operations/Distribution
William T. Pletcher

Operations Assistant
Donna M. White

Administrative Assistant
Georgiann Wright

Address all inquiries to:
Ortho Books
Chevron Chemical Company
Consumer Products Division
575 Market Street
San Francisco, CA 94105

First Printing in March, 1984

3 4 5 6 7 8 9
87 88 89

ISBN 0-89721-024-7

Library of Congress Catalog Card Number 83-62651

Chevron
Chevron Chemical Company
575 Market Street, San Francisco, CA 94105

Front Cover

Easy-maintenance red cedar shingles cover both the roof and the exterior walls of a seacoast home. Weathered to a silvery gray that matches adjacent buildings, the shingles let this new house fit gracefully into its turn-of-the-century community as well as its wooded site.

Page 1

On a home in the mountains, the blue of corrugated-metal roofing and metal-framed windows makes a lively contrast to warm-toned cedar-shingle siding. Once seen on residences only in Hawaii, and otherwise reserved for sheds and industrial buildings, metal roofing is now appearing on homes throughout the country as architects make use of newly available bright colors. Another view of this house appears on page 62.

Back Cover

Roofs and sidings on show:

Upper left: Metal panel roof with painted horizontal wood siding, page 6

Upper right: Cedar-shingle roof with stained redwood siding, page 4

Lower left: Asphalt shingles with stucco siding, page 16

Lower right: Corrugated-metal roof with cedar-shingle siding, page 62

Acknowledgments

Architects and Designers

Front cover: Huygens and DiMella, Boston, MA

Page 1; 62; back cover, lower right: David Wright, A.I.A., Nevada City, CA

Page 4; back cover, upper right: Rick Thompson, Sewickley, PA

Page 6; back cover, upper left: Kotas/Pantaleoni, San Francisco, CA

Page 7: Alexander Seidel, San Francisco, CA

Page 8: Robert A. M. Stern, New York, NY

Page 9: Construction Systems, Inc., Mountain View, CA

Additional Photography

Front cover: Steve Rosenthal

Page 4; back cover, upper right: California Redwood Association

Page 8: © Sepp Seitz/Woodfin Camp & Associates

Page 84: Alan Copeland

Special Thanks to:

Blair Abee

Pamela Allsebrook

Backen, Arrigoni & Ross, Inc.

Batey & Mack

James E. Caldwell Jr.

Julie Erreca

Fisher-Friedman Associates

JSW Architects

The Lawsons

John and Donna Rennick

Mr. and Mrs. D. Spolar

Don Tucker

Mr. Fred Vang and Ms. Martha Sage Vang

Consultants

Bob Beckstrom
Berkeley, CA

Pat Bowlin
American Plywood Association
Tacoma, WA

Gary Leucht
Novato, CA

Peter Johnson
California Redwood Association
Mill Valley, CA

Raymond Moholt
Western Wood Products Association
Portland, OR

Peter Rush
Asphalt Roofing Manufacturers Association
Washington, D.C.

Walt Tiberg
El Toro, CA

Franklin C. Welch
Red Cedar Shingle & Handsplit Shake Bureau
Bellevue, WA

Editorial Assistance

Beverley DeWitt
Karin Shakery

Copyediting and Proofreading

Editcetera
Berkeley, CA

Graphic Design Assistants

Mary Lynne Barbis

Susan Bard

Laurie Phillips

Illustration Assistant

Ronda Hildebrand

Typesetting

Vera Allen Composition
Castro Valley, CA

Color Separation

Color Tech
Redwood City, CA

HOW TO REPLACE & INSTALL ROOFS & SIDINGS

Roofs & Sidings: First Steps 5

Roof & Siding Terms *10*
How Roofs & Sidings Protect a Home *11*
Getting Started *12*
Safety Guidelines *14*

Roofs 17

Choosing the Right Roof *20*
Asphalt Shingle Roof *38*
Shake Roof *44*
Wood Shingle Roof *47*
Tile Roof *49*
Roll-Roofing *52*
Panel Roof *54*
Flashing *55*
Gutters & Downspouts *60*

Sidings 63

A Choice of Sidings *64*
Horizontal Wood Siding *69*
Vertical Wood Siding *72*
Panel Siding *74*
Shingle Siding *76*
Stucco Siding *78*

Maintenance & Repair 85

Caring for:
Roofs *86*
Gutters & Downspouts *90*
Siding *92*

Index *94*
Metric Chart *96*

ROOFS & SIDINGS: FIRST STEPS

Your home's roof and siding are its first line of defense against the elements. They also play a major role in how it looks. Whether you're building, remodeling, or just trying to keep your home in good repair, you'll want to give careful attention to the roof and siding.

Together, roofs and sidings form the protective exterior of your home. Applied layer by layer, they create a seamless series of overlapping materials that keep out rain, snow, hail, sleet, and howling winds. They also hold in heated or air-conditioned air. Whether you're building a new home or remodeling an old one, you'll want its roof and siding to be properly installed so that they can perform effectively.

Roofs and sidings also constitute most of the decorative elements of your home's exterior. Although you may have shutters, gingerbread trim on the eaves, or architectural embellishments such as a porch, the roof and the siding will be the main things a viewer notices. That means they should be appropriate to the style of the house, they should harmonize with one another, and they should be in good repair.

You'll find here all the help you need to install roofing or siding. If you're remodeling, this book will help you decide whether you have to take the old roof off first, or whether you can put the new one over the old—and then show you how to do both. If you are building, you'll find all the information you need on preparing the roof. Similarly, there are instructions for preparing a wall for new siding or for replacing old siding. There are details on the art of flashing a roof so it won't leak, and full instructions on how to install gutters and downspouts. And it is just as detailed in the application of siding, with discussions of the different types of siding and how to install them, from plywood sheets to shingles to stucco and more. In short, this book provides detailed instruction on how to provide your house with new roofing and siding, from start to finish.

This first chapter, "Roofs & Sidings: First Steps," introduces the subject, with photographs showcasing the beauty and variety of roofs and sidings, a discussion of the elements that go into the decision to install a roof or siding, a pictorial index to the terminology involved, an explanation of how roofs and sidings protect your home, and safety guidelines.

In the second chapter, "Roofs," you'll find in-depth information on choosing the right roof and planning its installation, followed by step-by-step instructions for installing the most common kinds of roofs. A large section is devoted to asphalt shingles, the most popular and the easiest for the do-it-yourselfer to install. It includes shingling techniques and patterns. Installation of shakes, wood shingles, tile, roll-roofing, and panel roofing is also fully explained. In addition, there are complete instructions for installing flashing, gutters, and downspouts.

"Sidings," the third chapter, covers everything you'll need to know to select and install the most popular sidings. It includes preparing the wall, installing jamb extenders, and using a story pole. There are full instructions for installing horizontal and vertical wood siding, plywood and hardboard panels, shingles, and stucco.

The final chapter, "Maintenance & Repair," provides the details of caring for roofs, sidings, gutters, and downspouts. There's information on fixing broken or bowed shingles, bubbles in tar-and-gravel roofs, and leaks in and around flashing. Maintenance of gutters and downspouts is explained, and there are instructions on how to repair cracked or rotted siding.

With the help of this book, then, you can have a roof and siding that effectively protect your home and contribute to its good looks as well.

Handsome roofing and siding materials, carefully chosen to compliment the style of a home and then carefully maintained over the years, add to both its visual appeal and its value. Here, rustic redwood siding coated with heavy-body stain is paired with cedar shingles on a traditionally styled New England home. Trim boards at the corners, around the windows, and along the edges of the dormer accent the horizontal siding.

A SAMPLING OF ROOFS & SIDINGS

Old and new combine in the hillside home above. Painted horizontal siding gives the multi-story house a traditional look, while the blue aluminum roof and the rounded windows emphasize its high-tech modernity. The roofing comes in 12-inch-wide panels that are "zipped" together. The redwood siding was installed with a special technique: at the corners, the boards were fitted into metal channels, eliminating the need for mitering corners or applying corner trim. Both roofing and siding were selected for their durability; the siding will need only periodic repainting, and the roof should be maintenance-free. These practical considerations join with design decisions to produce a thoroughly contemporary home with echoes of the familiar past.

Stucco serves the homeowner well. Properly applied, it forms a weather-tight seal on the home, lasts for years, and requires minimal maintenance; it can be applied in a wide range of finishes, and it takes color well. It also adapts to a multitude of architectural styles. With a red clay-tile roof, it characterizes the classic Spanish-style house of the Southwest and California. Combined with exposed timbers, it creates a Tudor effect. Countless bungalows are cloaked in stucco. Opposite, it demonstrates its adaptability to modern design. The simple, unadorned lines of this sleek house are made possible by the smooth-finished stucco. Both beautiful and versatile, stucco enhances all manner of homes.

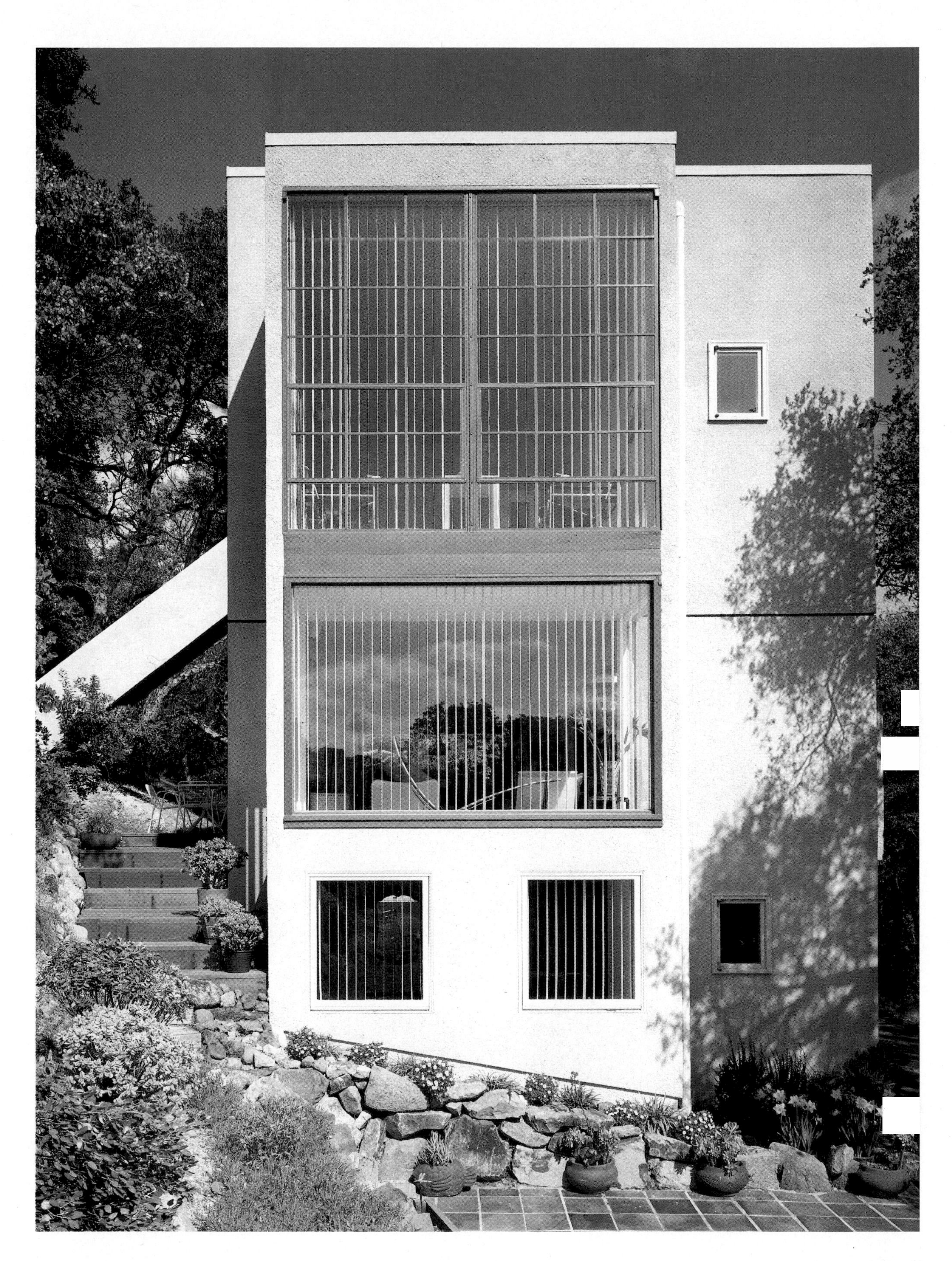

The seaside home at left is a modern classic. The "eyelid" dormer and the interrupted eave line above the door, the circle-in-a-square window, and the classical columns all mark departures from its basic form, that of an East Coast beach house of the 1910s and 1920s. Its heritage shows in the hipped roof line, railed veranda, simple, mostly symmetrical lines, and especially in the roof and siding, which recall the classic Shingle Style. The cedar shingles will stand up well to the rigors of coastal weather. They also provide a unifying texture, so that in spite of the variety and liveliness of the architecture, the house gives an impression of serene calm.

Vertical siding lends itself to the rectilinear lines of modern design. In the house shown above, doors and windows repeat the square and rectangular forms of the house itself, and the redwood siding's clean, straight lines are in keeping with the simplicity of these elements. The only ornamentation is narrow trim strips on the corners, along the eaves, and around the windows, neatly defining the shapes. Several coats of a clear sealant help preserve the warm tones typical of redwood siding; if left untreated, it would weather to a silvery gray.

ROOF & SIDING TERMS

Pictorial index of roofing and siding terminology

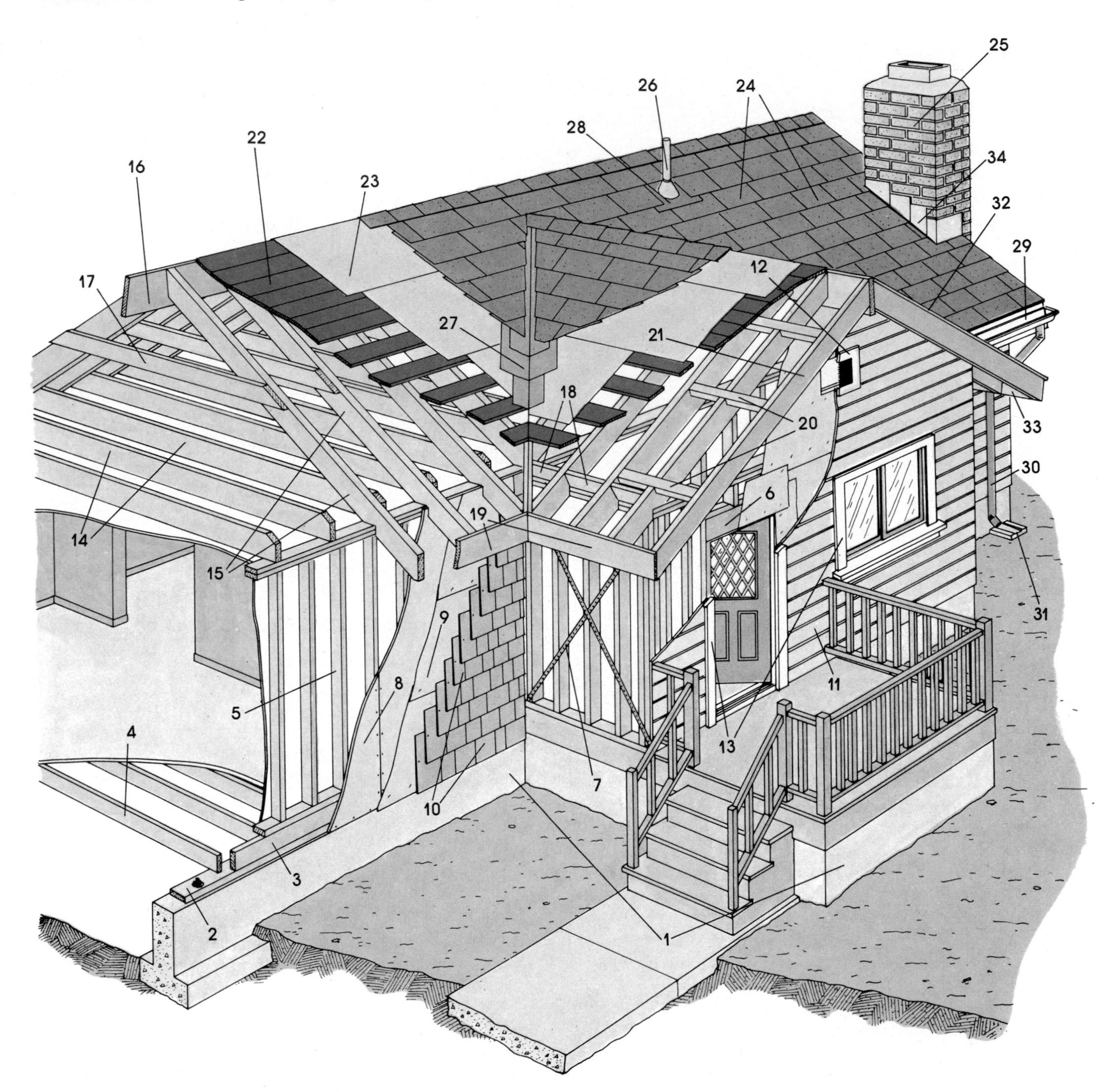

1. Foundation wall
2. Mudsill
3. Rim joist
4. Floor joist
5. Stud wall
6. Header
7. Wall bracing
8. Wall sheathing
9. Building paper
10. Shingles
11. Board siding
12. Attic vent
13. Trim
14. Ceiling joists
15. Rafters
16. Ridge board
17. Collar beam
18. Frieze blocks
19. Fascia
20. Outriggers
21. Barge rafter
22. Roof sheathing
23. Roofing felt
24. Shingles
25. Chimney
26. Vent pipe
27. Valley flashing
28. Vent pipe flashing
29. Gutter
30. Downspout
31. Splash block
32. Rake
33. Soffit
34. Chimney flashing

THE WELL-SEALED HOME

In a properly constructed building, weather protection begins at the ridge of the house—where ridge shingles overlap the roof shingles—and ends where the bottom edge of the siding overlaps the foundation wall. In the intervening area from ridge to foundation, each item of roof and siding overlaps the one below it, so flowing water cannot find its way into the house.

The illustration below shows how the various elements of roofing and siding provide this web of protection. Even those items that don't show, such as the roofing felt under the shingles and the building paper under the siding, must overlap.

The correct application of siding and roofing requires careful attention to detail. In new construction, the work proceeds from the bottom toward the top, with each succeeding layer protecting the previous one. From the illustration you can see, for instance, that a drip cap over a window is installed before the overlapping siding goes on, and the frieze blocks between rafters are placed to overlap the siding before the roof goes on. In doing a remodeling job, you must be careful to install or restore the pieces of your job in such a way as to maintain the protection built into the house.

This book will guide you in installing roofs and sidings so that your house is properly protected. For other projects, consult Ortho's books, *How to Replace & Install Doors & Windows, Basic Carpentry Techniques,* and *Basic Remodeling Techniques.*

The protected home

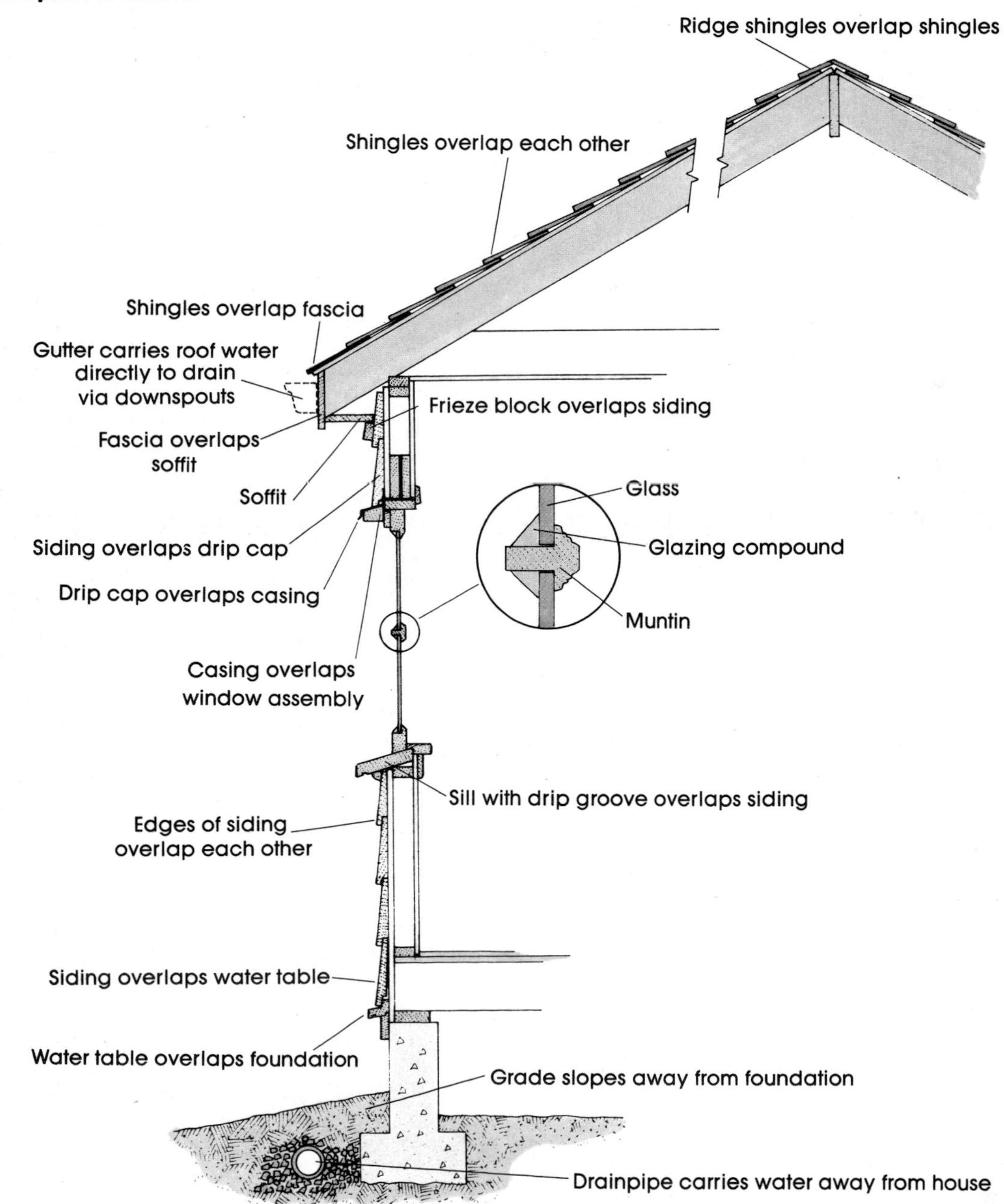

Looking for Trouble

If your roofing and siding concerns involve maintenance or remodeling of your house—as opposed to new construction—start by making a careful survey of your house from top to bottom.

Roof. First, stand back and look at your roof. If your house has asphalt shingles, as do the majority of houses in this country, look for dark patches, which indicate that the protective granules have worn away. These are weak spots in the roof. Look for curled shingles, caused by wind and temperature extremes. Water can work under those raised edges. On wood shake or shingle roofs, look for missing shingles. Once one is gone, the wind will start to pull others away. Watch for badly cupped or curled shingles, which could allow water to blow underneath.

Go up on the roof (see Safety Guidelines, pages 14–15) and look at the flashing around vent pipes, chimneys, or the conjunction of two walls. Cracks or gaps in caulking material are a possible leak source. See that the flashing on the chimney is still firmly embedded in the mortar between the bricks. Loose mortar will allow water to get behind the flashing.

Carefully check the valleys on the roof—these can be a source of trouble. See that the valley material is undamaged and in good condition. In addition, make sure the valleys are free of debris. Leaves or pine needles can dam up water and cause it to run out the sides of the valley and under the shingles.

On flat roofs, a low spot where water collects is potential trouble. Look for bubbles in a tar and gravel roof, which often indicate moisture has worked its way under the roofing felt. Fix any bubble (see page 87). Otherwise, "don't fix anything that ain't broke."

From a ladder, check the eaves of the roof. This is a particularly common area for rot to set in because water can be driven up under the shingles through rain or freezing action. Push the tip of a screwdriver into the wood on the edge of the eaves. Anyplace where it penetrates much beyond 1/4 inch probably has wood rot. Fortunately, any rot in these areas generally does not extend far up the roof and can be repaired without much difficulty when a new roof is applied. Have a good look at the gutters while you are up there. They should be clean, without sags, and firmly in place.

Now take a good flashlight and go into the attic. Carefully look over each rafter. Even if you haven't seen a leak in the house, water could be leaking through the roof, running down a rafter, and dropping between the siding and the inside wall, a situation you can spot by telltale water stains on the rafter. Use your screwdriver to poke any suspicious-looking spots for possible rot.

Siding. Next, have a careful look at the siding. It may have problems, but generally there will be fewer than you might find on a roof because it is less subject to sun and storm damage.

Look at the paint job first. If you see extensive bubbling, cracking, or peeling, it may mean that you have insufficient vapor barrier protection in the house. This can happen if your house is not insulated, or if the insulation does not have a vapor barrier on it, such as the aluminum foil on insulation batts. The vapor barrier is designed to stop the moisture in your house—generated by such activities as breathing, washing dishes, doing laundry, and bathing—from moving through the walls where it may condense on meeting cold outside air. The vapor can move through the house sheathing and literally blow the paint off it. One solution, short of tearing off the inside walls and installing vapor barriers, is to add vents inside the house and install dehumidifiers.

Look for any cracks in the exterior sheathing. Water can work its way through a crack to the inside of the wall and cause wood rot.

Check the trim around the doors and windows and at the corners of the building. It should be firmly in place and well caulked. There should be a tight seal where siding meets chimney masonry.

Spotting a leak

How to Find a Leak

Leaks in the roof may be difficult to find. Rarely, for instance, does a leak in the ceiling come from a spot directly above in the roof. Instead, it may have traveled a considerable distance along the roof deck or a rafter before appearing.

When a leak does develop, go outside and see if you can spot a loose or missing shingle. If there is nothing obviously wrong on the roof, take a flashlight into the attic. Leaks are commonly found around vent pipes, chimneys, skylights, and valley flashing. With the exception of valley flashing, these are areas where the roof deck has been cut through and then sealed. The seal, or flashing material, is often the culprit. In the case of valley flashing, the leak is generally caused by an obstruction in the valley that causes water to pool behind it, or by poor workmanship.

Starting from above the leak in the ceiling, look for water running down a rafter. Follow that up the rafter. If it disappears, it is most likely running along the side or top of the rafter. You may have to pull away insulation to trace the water. If the insulation is wet, remove that section by cutting above and below the damp area with a knife. Dry it, then staple it back up after the leak is fixed. Wet insulation is an invitation to mildew and rot.

If you find a spot where the water appears to be coming through the roof deck, push the end of a straightened coat hanger through it. You will then be able to spot the leak site when you get on the roof.

Do not go up on a roof in a rainstorm. Just put out a pan to catch the drip and wait until the roof is dry. Working on a slippery, wet roof can be bad for your health.

When you get on the roof, use a hose to flood the suspected area with water while someone in the attic watches for the drip. Begin watering the area below where you think the leak might be and slowly work your way up. This will help pinpoint the leak.

Another way to spot a leak is to make the attic dark and then look for a ray of light, which indicates a hole. Search carefully around vent pipes and chimneys.

Planning a New Roof or Siding

There are a number of factors to take into consideration when planning to install a new roof or siding, among them choosing an appropriate style, estimating the cost, and estimating how much time the project will take.

What Style?

Installation of a new roof or siding provides an opportunity to change the way your house looks. If your house has a particular style, you'll probably want to keep that look, or perhaps even carry it further—replacing asphalt shingles with shakes, for instance, to compliment a Colonial. Or you may want to modernize an old-fashioned-looking house. Whatever your preference, be sure to read through the rest of this book before making your decision. The pitch of your roof may limit your choices; the application of some materials—such as a slate roof or aluminum siding—is beyond the scope of the do-it-yourselfer, and thus of this book, because the work requires professional skills.

Once you know the practical limitations on the choice of style, give yourself some time to consider the variety of options remaining. The photographs on pages 6–9 can help you focus on the possibilities. Lumberyards and roofing centers can show you some ideas, and manufacturers are happy to mail brochures that show what their products can do. You might consider, for instance, adding antique molding to the trim on your home, or putting fancy-cut shingle siding on dormers.

If you plan on radically altering your house's appearance, consider how it will fit into the neighborhood when you are finished. Ordinarily, a new roof or new siding won't raise any eyebrows, but a radical departure from the norm—say, a shimmering aluminum panel roof in a line of houses covered with sedate dark shingles—may make you an unwelcome neighbor.

How Much Will It Cost?

New roofing and siding are expensive projects, but ones that sooner or later must be done. Many roofs don't last much beyond 25 years. Siding, depending on the quality of material and paint, may hold up longer, but it won't last forever.

While these jobs may be expensive, they will cost you only half as much if you do the work yourself. It takes some skill and hard work, but you can do it—and do it right—with a little patience. Remember the motto of a wise old carpenter: "The lazy man works twice as hard."

How Much Time Will It Take?

Fair warning is needed here: never underestimate the time needed for roofing and siding projects. Sit down and make some careful estimates, using the instructions in this book as a guide to the magnitude of the job. Whatever time you come up with, double it—and even then have contingency plans for unexpected delays.

For openers, it's a Roofer's Law that no matter what time of the year you take off the old roof, you can expect it to rain. Even if it hasn't rained for three months, count on showers when the old roof is removed. You can protect your house (see page 26), but you won't be able to work until the rain stops.

More possible delays: you find rotted roof deck boards under the old roof that must be replaced; a rafter has to be fixed; you find that you need to replace the gutters too; or the flashing around the chimney is worn out and must be replaced.

Putting on new siding is not as subject to unexpected delays—the eaves will provide considerable protection in inclement weather, for instance—but it too can be a time-consuming job.

This list of potential calamities is not meant to discourage you, only to make you aware of the possibilities. A job with no adverse surprises is a pleasure, and a rare one at that. To save time, you may need help. Your spouse probably has something better to do, but there may be a youngster in the neighborhood willing to earn money. Look in the service directory of the local paper or call the high school. If worse comes to worse, pay your own kids: their willingness to work for real money might surprise you.

Codes and Permits

Virtually every house built or remodeled in this country must adhere to one of the national model building codes, or a local one. Such codes are designed to ensure safe construction practices. When you undertake any construction project, you usually must have a permit issued by your local building inspector, but installing a new roof or siding on an existing house may not require one. To find out, you will have to call your inspector and inquire. In roofing, the concern is whether the new roof will add too much weight to the rafters. For siding, there will be such questions as whether, when you remove old siding and put on the new, the walls will still be structurally braced. Local codes vary so widely that you will have to discuss your situation with the building inspector. The permits are not particularly expensive, so don't try to do without one—this could get you fined.

SAFETY GUIDELINES

Working on a steep roof or standing on a high ladder to do a siding job has one obvious risk: it's a long way to the ground. Working without proper precautions or being careless on the job is an invitation to trouble. But by staying alert and using proper safety equipment and techniques, you can make your job trouble-free.

The Worker

Attitude

If you are quite nervous about being on a roof or high on a ladder, don't force yourself to do it. Many people are disconcerted by heights. Telling yourself that you are going to do it anyway may be courting disaster. Work your way up slowly—over a few days, perhaps—to see if you become accustomed to the height. If you still feel afraid, get a qualified person to do the job.

If the work starts going badly and your frustration rises, take a break. Getting angry and hurrying the job only invites more mistakes. The job should be relatively pleasant. After all, you're doing it on your own time.

Clothing

On a roof or a ladder, wear soft-soled shoes such as tennis shoes to minimize the chances of slipping. Such shoes will also do less damage to roofing material than heavy work boots. This is particularly true with asphalt shingles, which will turn soft in a hot afternoon sun. Hard-soled shoes or even direct knee pressure can break the protective layer or cause a tear in the shingle.

Shirt and pants should be loose and non-binding so you can move about easily. Roofing and siding can be long jobs, so stay comfortable.

Wear protective eye gear when chipping concrete, or when using a power saw to cut metal, fiberglass, or masonry.

The Equipment

Ladders

Correct use of ladders is of paramount importance. Many people use ladders rather nonchalantly, and only start thinking about ladder safety after a near mishap.

Before using the ladder, check that all the rungs are solid. See that there are no loose screws or rivets.

On extension ladders, make sure the pulleys work smoothly. Double-check the locking devices. Sight along the ladder to see that it is not bent, which could cause it to collapse when weight is placed above the bend.

To raise a long ladder, have a friend brace the bottom with his or her feet while you walk it up over your head from the other end.

When a ladder is positioned against a house, set the ladder base away from the house by a distance equal to one-fourth of the working height of the ladder.

Ladder feet should be set on firm and level ground. If one leg must be shimmed to keep it level, use a plywood scrap. Don't build up a series of blocks.

Ladders placed on a smooth concrete surface may slip. Counter this by gluing rubber or indoor/outdoor carpeting to the bottom of the ladder feet.

Using a ladder safely

Stepladder-and-plank scaffold

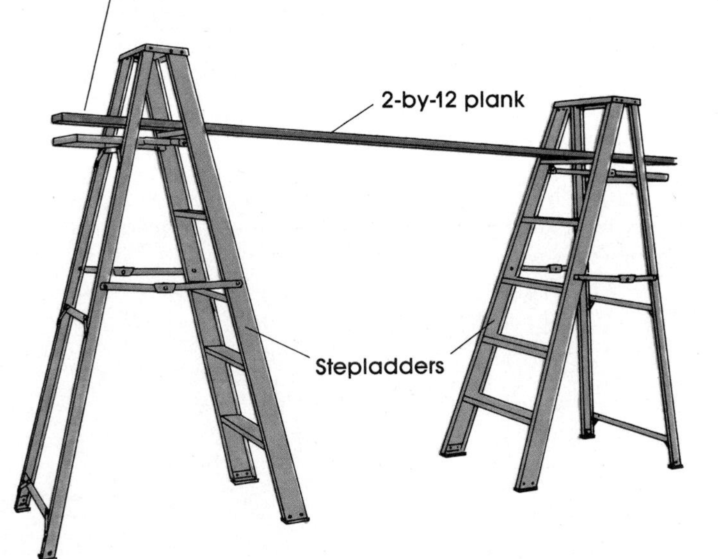

When you raise a metal ladder, make sure it does not come in contact with overhead power lines. New fiberglass extension ladders eliminate the chance of getting a serious shock.

When working with a long extension ladder, keep the ladder from slipping to the side by securely tying the top to a gutter support strap.

When working on a ladder, don't lean out in an attempt to reach beyond arm's length. Keep your body inside the ladder rails.

Ladders and Scaffolding

Two ladders with a plank between them make an effective scaffold that greatly speeds up a siding job.

A 2 by 12 makes an effective plank. Make sure each end extends 1½ feet beyond the support, since your weight will cause the plank to bend and pull the ends away from the supports. Aluminum planks can be rented to span a longer distance with less deflection. To create a scaffold, you might want to rent ladder jacks, which fit over the rungs (see illustration).

Roof Rules

Never go up on the roof when it is raining, when the roof is wet, or if a lightning storm is imminent.

Wear soft-soled shoes, as noted above, to minimize the chances of slipping or damaging the roof.

Keep the roof clear of debris. Stepping on a scrap of roofing felt can send you flying in a split second. Watch out for loose shingles or tiles, moss, and wet leaves.

When removing an old roof, work from the top down and keep the roof clean by sweeping periodically. Debris should be dumped in a container directly from the roof, if possible. When throwing things from the roof edge, be careful not to overbalance.

The area where you are dumping debris should be roped off so no one strays into this free-fire zone.

Materials stacked on the roof should be dispersed evenly to spread the load. If the roof is too steep to keep shingle bundles from sliding, nail roofing jacks near the ridge and install planks to keep the bundles in place.

If your roof has a pitch of 5 in 12 or less (see page 21 for calculating roof pitch), it should be fairly easy to work on. But on roofs with a steeper pitch, you may need to take some extra precautions. Working next to the eaves when applying the first few courses on a steep roof is awkward and dangerous. Use ladders and a scaffold plank if possible when applying the first courses, to provide enough room for you to sit on the roof as you shingle.

If scaffolding is not possible, tie yourself to a rope that goes over the ridge and is well secured on the other side to a tree, porch upright, or the like.

Use ladder jacks or a roofer's seat as described on page 45 as you continue up steeply pitched roofs.

Stay alert and don't take any chances. Otherwise, make sure your insurance is all paid up.

Ladder jacks

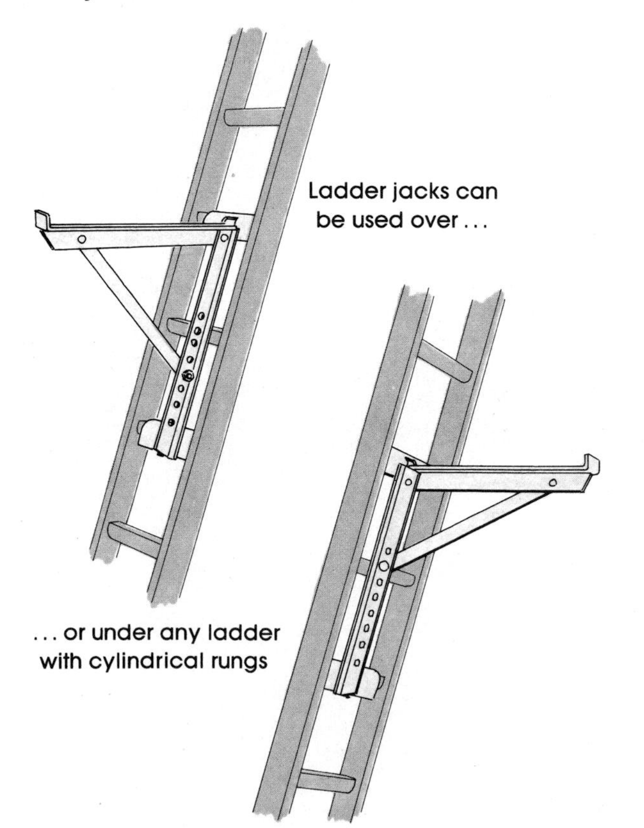

Scaffolding

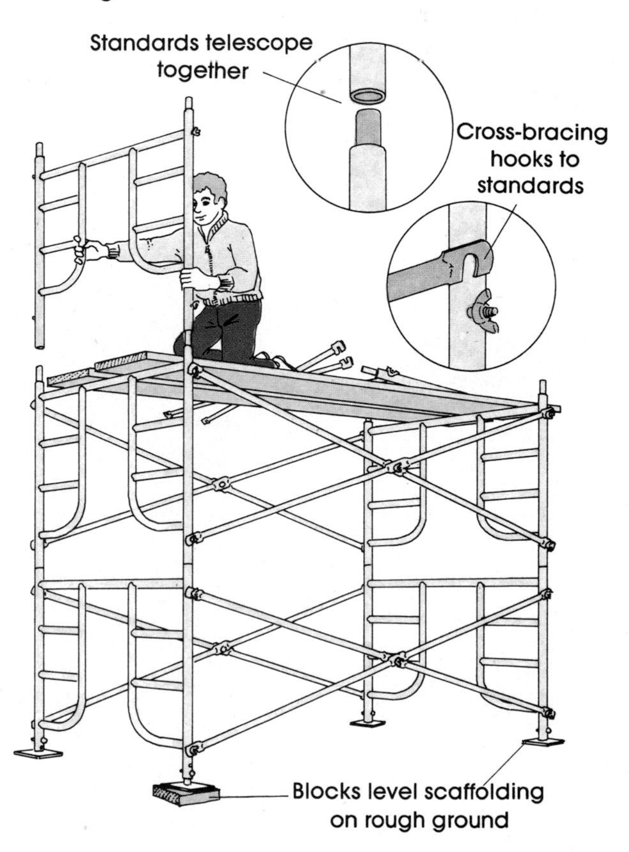

ROOFS

There's a wide variety of roofing materials, many of which you can install yourself. In this chapter, you'll find everything you need to know—selecting the right material; preparing the surface; applying asphalt and wood shingles, metal and fiberglass panels, shakes, tile, and roll-roofing. Special sections on installing flashing, gutters, and downspouts.

Presumably, you bought this book because you are going to apply new roofing or siding—or both—to your house and you want to do it yourself. You can almost assuredly put your own siding on by following the directions in this book. However, when it comes to a roof, you need to take a number of things into consideration.

One of the first questions to ask yourself is how fast you can work.

If you are removing an old roof, or if you are working on new construction, you must get the roof on quickly. Don't even imagine what your house would look like after a thundershower put 2 inches of rain in the attic.

In estimating job time, you should allow two hours to put down the first square (100 square feet) of shingles. The second square will go a little faster, and by the second day you might be putting down one square per hour. By comparison, a professional can lay two squares or more an hour.

Another consideration is how hard it is to work on your roof. Any pitch above 8 in 12 (see page 21 for calculating roof pitch) is difficult. If you have such a steep roof, particularly one with many dormers, hips, and valleys, then you should consider hiring a professional.

Consider also your willingness to do the physical labor involved in roofing. Heavy materials have to be toted around. You'll be spending many hours hammering. Be sure you're ready to undertake the work; otherwise, you may find yourself calling in a professional to finish your half-done job.

Reasonably priced, easy to apply, and available in a wide variety of colors and styles, asphalt shingles cover more roofs in this country than any other roofing material. At left, dark-brown rustic-style shingles make a fitting roof for an English Tudor-style house.

Certain roofs—tar and gravel, slate, and aluminum shingles, for instance—should be done professionally. If one of these is your choice, or if you're hesitating about installing one of the other roofs yourself, you can call in a contractor.

Working With a Contractor

You can hire a contractor to do your entire roof, or to help with some parts of the job.

Contractors are licensed by the state. They must include their license number on any contract, and you can go to the state contractor's licensing board with any complaint you might have.

Get the names of several professionals and ask them to bid on your job. Follow up on their personal recommendations to learn if their work has been satisfactory. When you select someone, ask that they do the following:

- ☐ guarantee their work in writing for a specified time period.
- ☐ specify the materials to be used in the job.
- ☐ specify exactly what jobs they will perform.
- ☐ specify when the job will be completed, and how they will compensate you if it is not completed in time.
- ☐ agree that you give them no more than one-quarter of the fee in advance, with other increments to follow at specified periods.
- ☐ specify that any subcontractors they hire will be paid by them.

You may wish to try a specialist rather than a contractor. A specialist is essentially a skilled worker who is not licensed; any dispute with a specialist would have to be worked out with that individual alone. Many do work that is just as good as a contractor's; they may also work for less than a contractor. However, with a specialist, you may have to pay a workman's compensation insurance fee to protect yourself in the event of an on-the-job injury. Check with your own insurance agent on this matter.

ROOFING TOOLS

Fortunately, roofing does not take a lot of expensive or unusual tools. You can probably get the job done with what you have in your workshop now. But, as with most jobs, roofing is much more enjoyable if you have the right tools. Remember that roofing is more than just nailing down shingles. Maybe the old roof has to be removed first. Then there are drip edges, flashing, gutters, and downspouts to be put on. Here are some tools useful in a roofing job:

Tin snips. These are needed to cut flashing and metal drip edges to length.

Tool belt. A standard carpenter's tool belt has four pockets for holding different sized nails and several loops for holding such things as tape measure and hammer. A tool belt is a time-saver because all your equipment is readily at hand, and moves when you do.

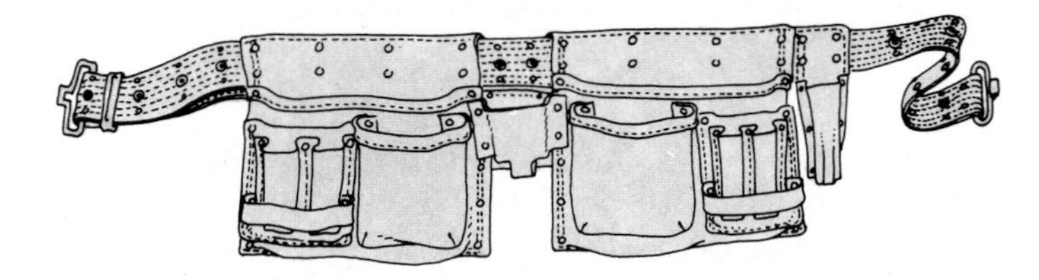

Chalkline. This tool has a string wound in a container full of chalk. When stretched tight and snapped, the chalk-coated string leaves a straight line over a distance of up to 100 feet. Chalklines are snapped periodically to keep shingle courses properly aligned.

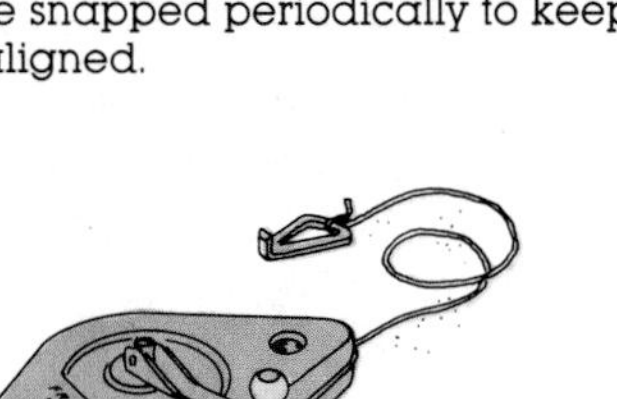

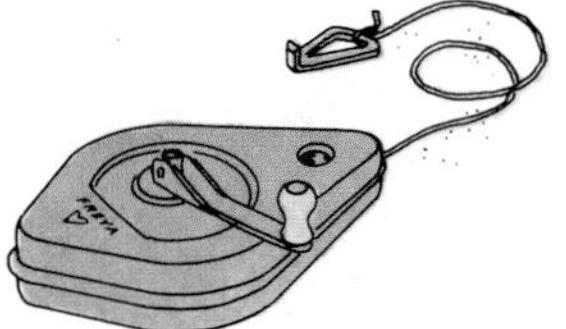

Pry bar. This heavy metal bar is useful when removing nails and wood shakes or shingles from a roof.

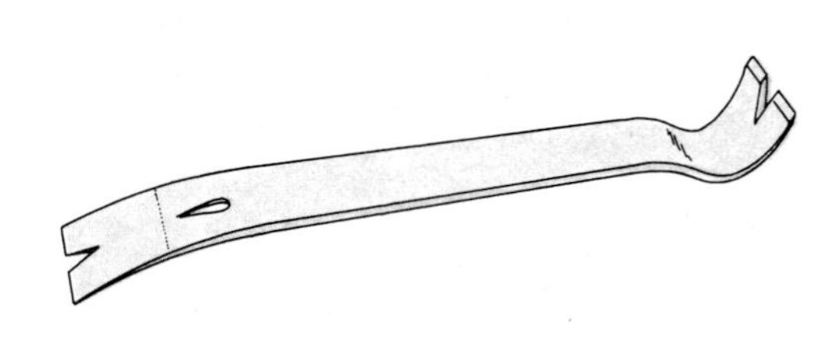

Tape measure. A retractable steel tape at least 16 feet long (preferably 25 feet) is needed for installing a neat and professional-looking roof.

Chisel and saw. For removing and repairing any problem areas in the roof sheathing.

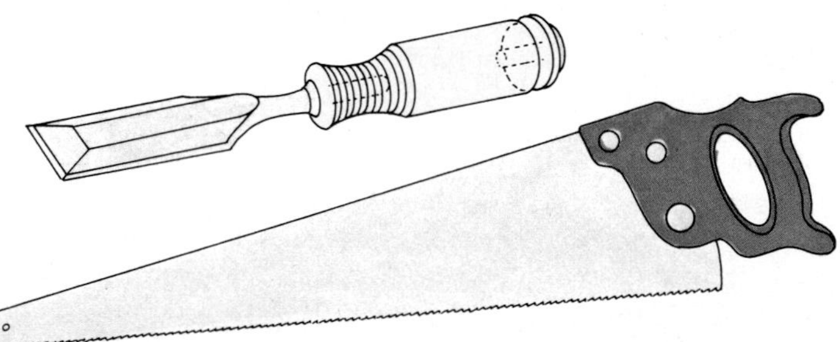

Caulking gun. For applying beads of caulk around flashing and at the edge of shingles in valleys.

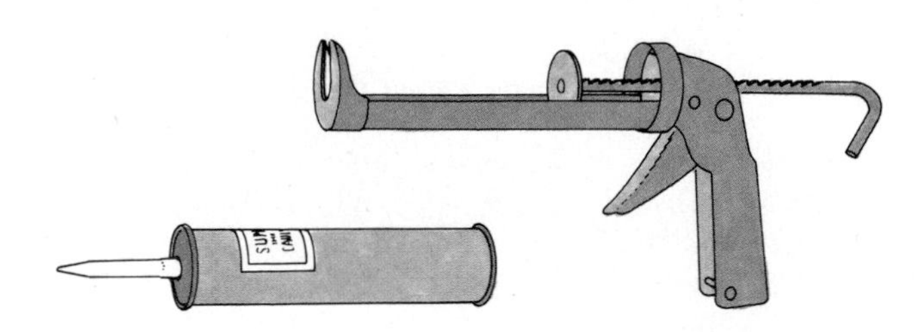

Putty knife. A putty knife can be used to apply roofing cement around vents. However, the tail of an old wood shingle works just as well and can be thrown away when you're finished.

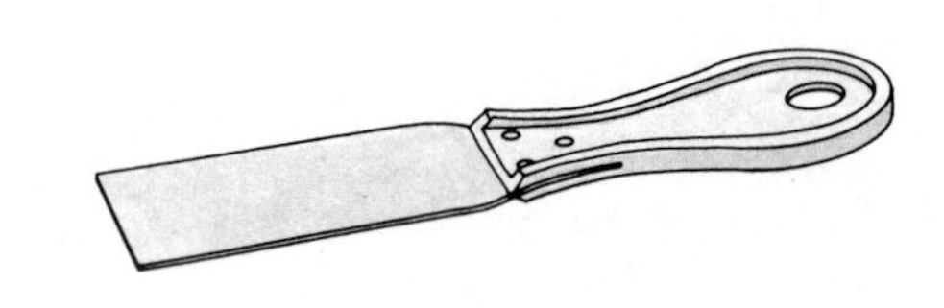

Roofer's hatchet. Although you can get by with a hammer, the specially designed roofer's hatchet is well worth the cost. There are two basic types available. A style commonly found in hardware stores has a combination hammer and hatchet for splitting wood shingles or shakes. The hatchet side has a series of holes ½ inch apart that accept a knurled knob. This creates a shingle exposure guide on the hatchet. Measure from the hammer face to the hole of your choice, depending on how much shingle exposure you want (see illustration), and screw in the knob. Professional roofers use a similar hatchet that has a small replaceable knife blade on the hatchet end for cutting asphalt shingles. Dull blades can be quickly replaced.

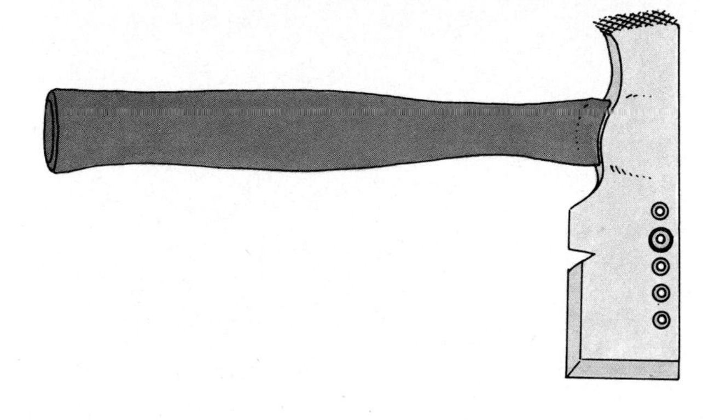

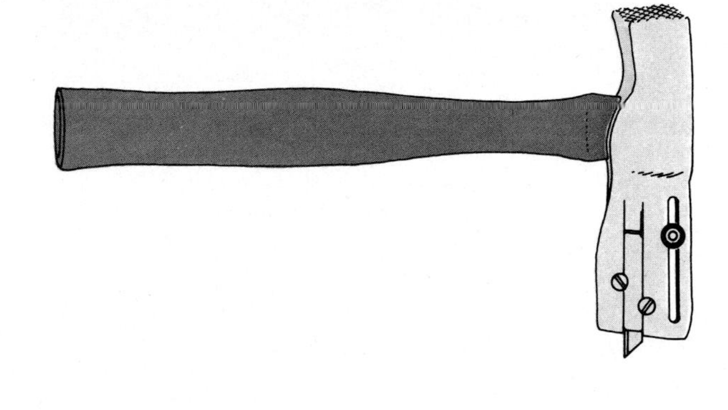

Utility knife. Choose one that contains several blades that can also be reversed. This knife is most commonly used when cutting asphalt shingles.

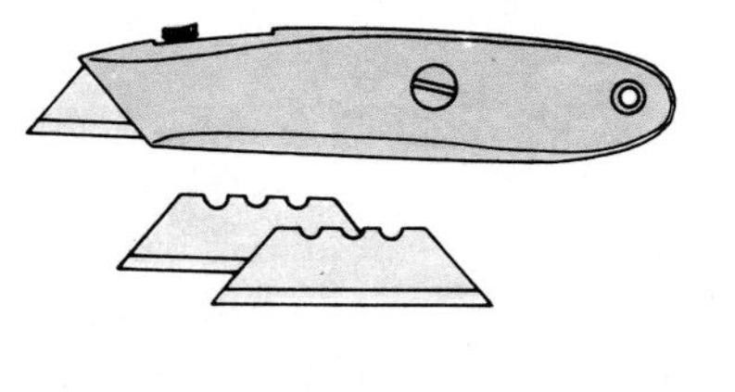

Roof jacks. These are used on steep or slippery roofs to prevent falling (see below). They can be rented.

Using Roofing Jacks

Roofing jacks are lengths of metal strap that hold a board to help support you on a steep slope. Most of your weight stays on the roof, but the board keeps you in place. You can also walk back and forth on the roof with part of your weight on the board. Jacks can be used on all types of roofing.

One end of the roofing jack has notches so it can be nailed to the roof; the other end has an angle to hold the edge of a board. Roofing jacks come in different sizes and shapes; the one that holds a 2 by 4 on edge is excellent. Do not use a board longer than 10 feet; you don't want to risk having it break under you.

Jacks should always be firmly nailed through the decking and into a rafter. Always put the nails where they will be covered by a course of shingles.

If you have never seen roofing jacks in use, you may wonder how it is possible to nail a jack to the roof, cover it with several courses of shingles, then remove it and move up. The trick is in the shape of the top of the jack. When you want to remove the jack, tap the bottom with the hammer until it slides up free of the nails, then pull to one side and slide it out.

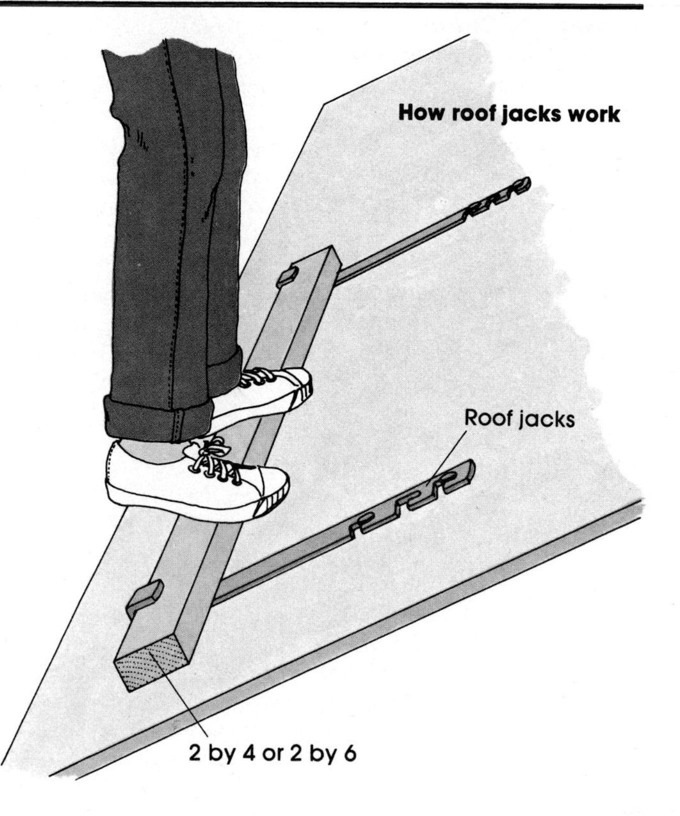

CHOOSING THE RIGHT ROOF

Roof Styles

None of the six basic roof styles illustrated here is particularly "better" than another, but some are better suited to certain parts of the country. Flat roofs are common in much of the Southwest, where there is relatively little rain or snow. However, such a roof would not be practical in states that often endure heavy snowfalls, or even in high elevations in the Southwest. There you commonly find steep gable or hip roofs that readily shed snow.

Roof style dictates roofing material to some degree. Flat roofs, for example, are almost always covered with tar and gravel, making what is called a built-up roof. A nearly flat roof might also be covered with asphalt roll-roofing, which is inexpensive but unlikely to last beyond ten years. Built-up roofs are widely used on roofs with up to a 3 in 12 pitch (see opposite for calculating roof pitch), but not greater than that because the hot tar runs off when it is applied.

With some special preparation, asphalt shingles can be put on roofs with as little as a 2 in 12 pitch. However, a steeper (4 in 12) slope is greatly preferred in order to shed water more quickly. Shake and shingle roofs should also have at least a 4 in 12 pitch to efficiently shed rainwater.

The slope of your roof will also be a consideration in your decision to do the work yourself or hire a contractor. If the roof is very steep, or is two or more stories high and has numerous dormers or a turret or two, you had better get a contractor. But if you have a more standard roof, you can probably do it yourself.

Types of Roofing Materials

Asphalt Shingles

It is estimated that more than 70 percent of the houses in this country have asphalt shingles, sometimes called composition shingles.

Asphalt shingles have a central core made from cellulose fibers or fiberglass that is coated with asphalt on both sides and topped with a protective mineral aggregate.

Of all the different types of shingles available, the easiest for the do-it-yourselfer to apply—and the most widely sold—are three-tab asphalt shingles (see illustration, page 22). They are 12 inches wide and 36 inches long. (Increasingly, shingles are being made in metric measurements. Be sure you know what you have purchased, and check the manufacturer's installation instructions. The instructions in this book use inches and feet. A metric conversion chart appears on page 96.) Three-tab shingles come in different qualities, judged primarily by their weight per square (the quantity needed to cover 100 square feet). The heavier the shingle, the longer it is guaranteed to last. Standard shingle weights run about 215 to 300 pounds per square and are designed to last 15 to 20 years. Asphalt shingles can also be two-tab, one-tab, or interlocking, but these variations are less available in stores nowadays.

Asphalt shingles are now commonly made with self-

Six types of roofs

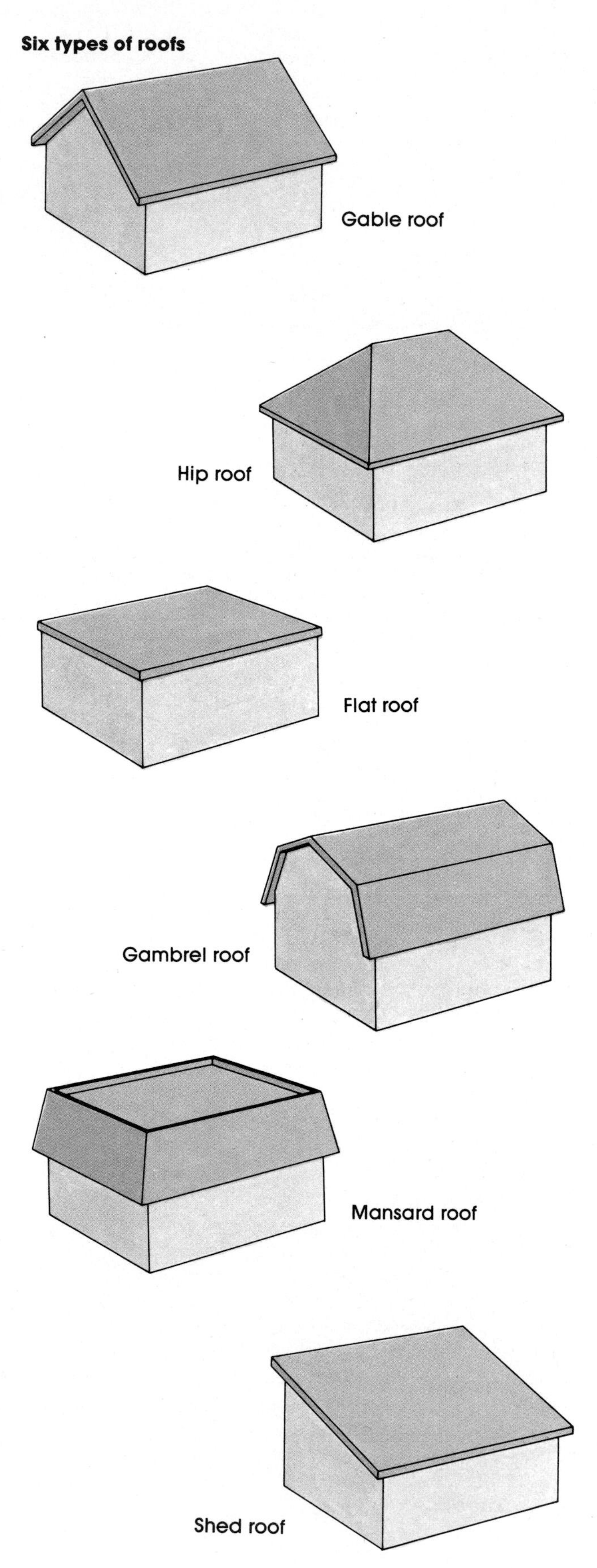

Determining Roof Pitch

The slope, or pitch, of your roof is important in deciding what type of roofing material you can put on it. There are two basic ways to determine roof pitch.

If you can readily get on your roof, you can easily calculate the pitch with a straight board, a bubble level, and a ruler. Place the board, which need be only about 3 feet long, on the roof. Check that it closely conforms to the slope of the roof and is not pushed out of line by a rock or a warped shingle. Measure and mark off 12 inches on the level. Place one end of the level near the top of the board and raise the other end of the level until the bubbles are centered. Holding the ruler so it is straight up and down, measure the distance from the top of the board to the bottom of the level at the 12-inch mark. The ratio of that distance to 12 inches tells you the roof pitch. If the distance is 6 inches, that means the roof rises 6 inches for every 12 inches of horizontal run. Pitches normally range from 1 in 12 (nearly flat) to 12 in 12 (45-degree angle typical of an A-frame) and normally do not come in half-inch increments.

Another way to check the roof pitch is with a tool called a Squangle. This adjustable device, available in most hardware stores, is used in laying out and cutting rafters. Just place it against an exposed rafter end or a block placed against the fascia board, adjust until it lines up with the angle of the roof, and read the scale.

Two methods of determining roof pitch

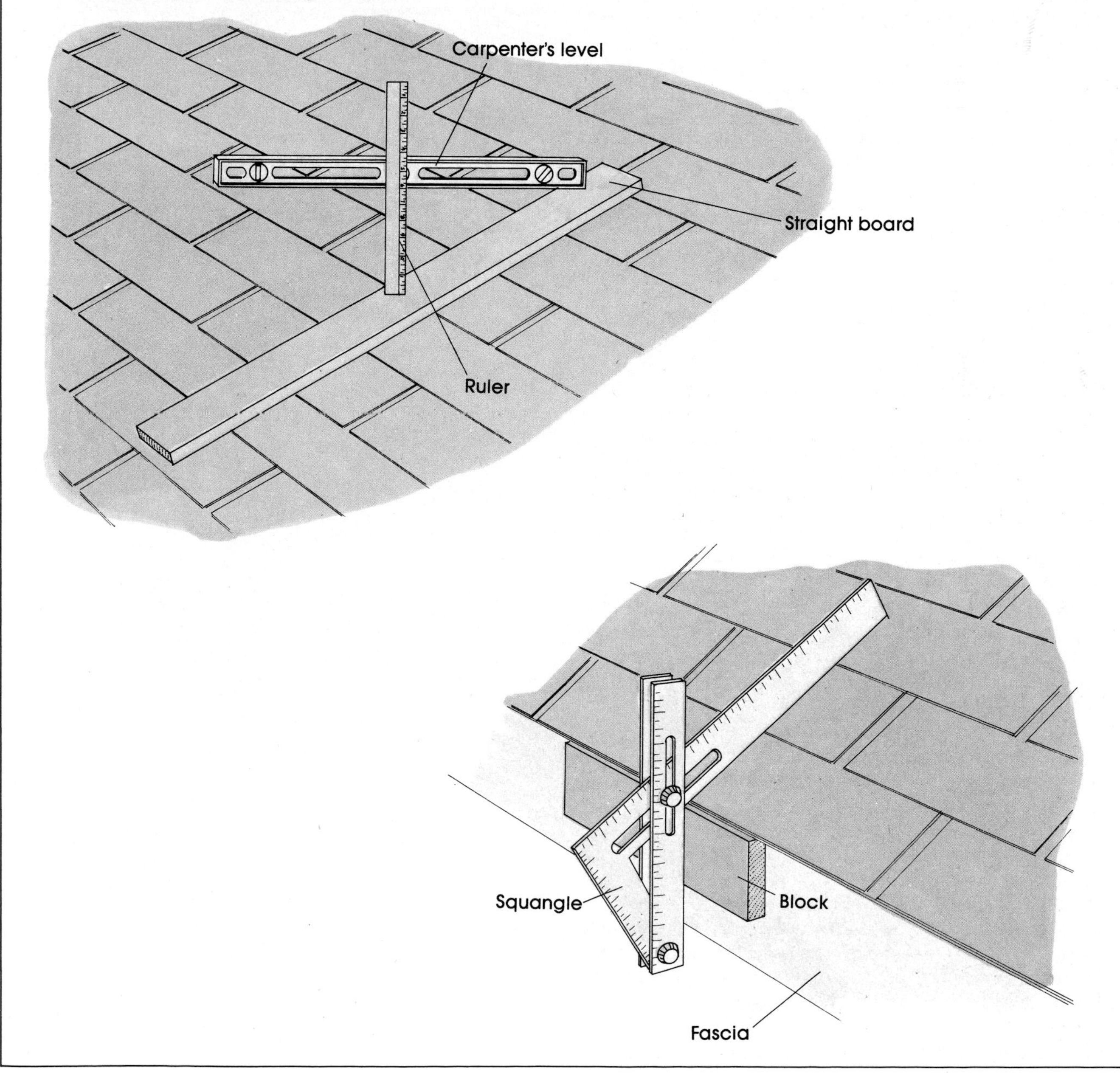

sealing strips on the surface of the shingle to prevent the wind from blowing them loose. After the shingle is applied, the heat of the sun bonds the strip, adhering the shingle to the one above.

Ideally, asphalt shingles should be laid when the temperature is between 40 and 85 degrees F. In cold weather they should be stored in a warm area prior to installation.

Wood Shingles and Shakes

Shingles are smaller and lighter than shakes and commonly are sawn on one or both sides, which gives them their typical shape. Shakes are usually split by hand, which helps account for their higher cost. Both shingles and shakes are mostly cut from western red cedar.

Shakes and shingles are graded by number—1 (the best), 2, and 3. For roofs, use only number 1. Cut from heartwood, it is highly resistant to rot and free of knots. For siding you can use a number 2 grade, which has more sapwood (outer part of tree) and some knots. Number 3 has even more sapwood and knots.

Shingles come in lengths of 16 inches, 18 inches, and 24 inches. They are sold in bundles, with four bundles to the square, which will cover 100 square feet at 5-inch exposure.

Shakes are sold in bundles of 18-inch or 24-inch lengths, and in either a heavy- or medium-weight grade. A five-bundle square of 24-inch shakes will cover 100 square feet at 10-inch exposure.

Both shakes and shingles are commonly applied over spaced 1 by 4s or 1 by 6s, rather than over a solidly sheathed roof, to allow air circulation (see pages 44–48). This is especially true with shingles, which, because they are machine sawn, fit tightly and smoothly together. Roofing felt (see page 33) is never put under shingles, but is always placed under shakes, since their irregular surface allows air to circulate.

Neither shakes nor shingles should be used on roofs with less than a 4 in 12 pitch since they will not readily shed water, and wind can blow rain under them.

Roll-roofing

Roll-roofing is essentially the same material as asphalt shingles. However, it does not last nearly as long, because there is only one layer over a roof, while the overlapping of shingles means there are acutally three layers of material on the roof.

There are two basic types of asphalt roll-roofing. Roofing felt, an asphalt-impregnated paper that is used as a protective underlayment for many different types of roofs, is also a type of roll-roofing.

Mineral-surface rolls. These are similar in color and appearance to composition shingles. Mineral-surface roofing is commonly called 90-pound felt because one roll, which covers one square, weighs 90 pounds. It comes in rolls 36 inches wide and 36 feet long, which totals 108 square feet; but because of overlaps, the actual coverage is 100 square feet. It can also be used as flashing and is commonly put in valleys under metal flashing.

Selvage-edge rolls. Also called split-sheet roofing, these

Roofing materials

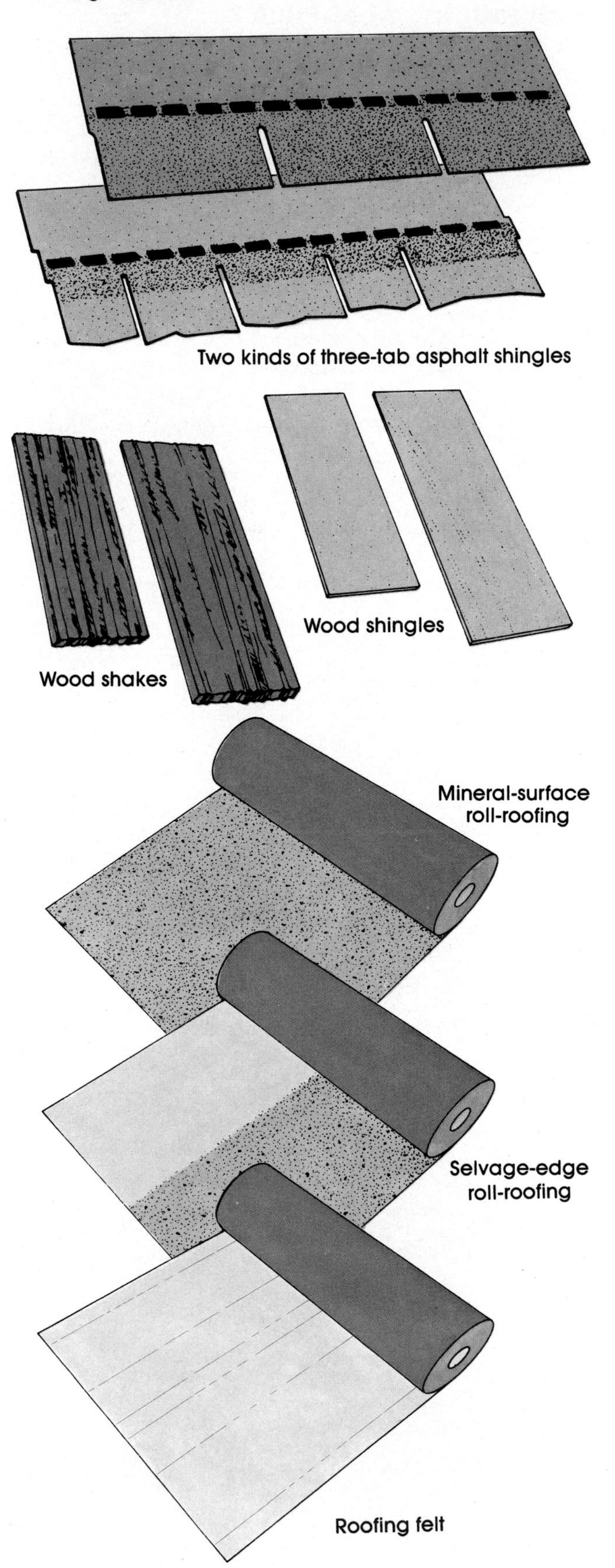

Roofing materials

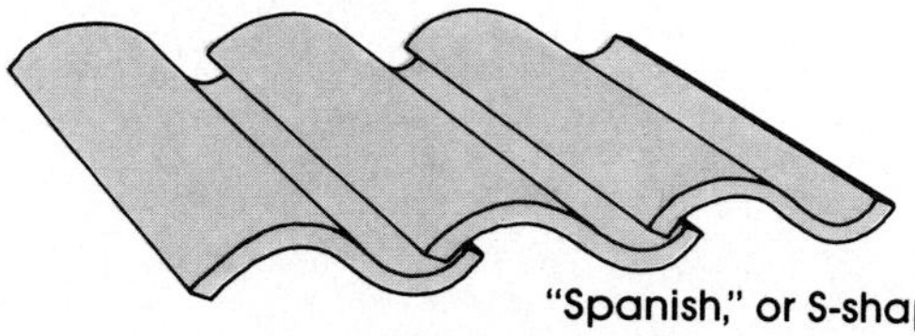

"Spanish," or S-shaped, clay tile

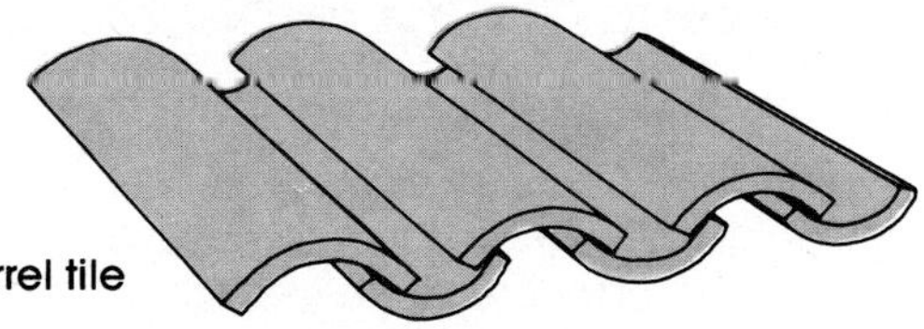

Clay barrel tile

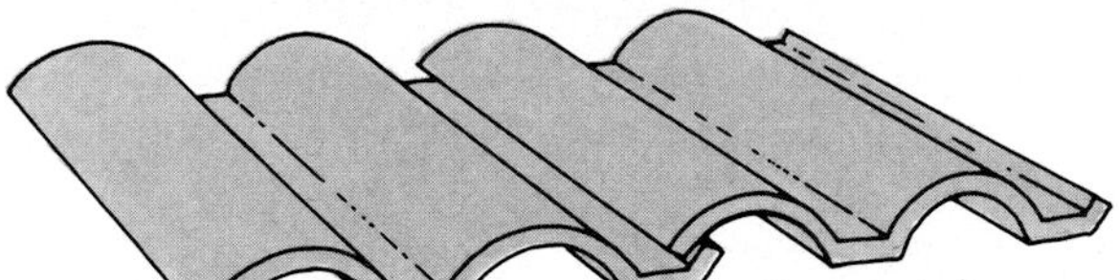

Concrete barrel tile

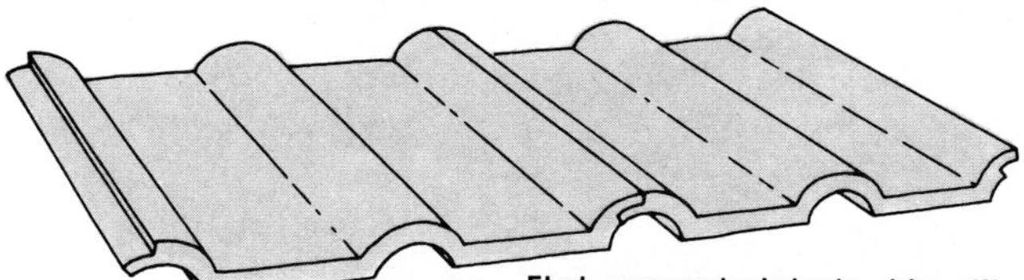

Flat concrete interlocking tile

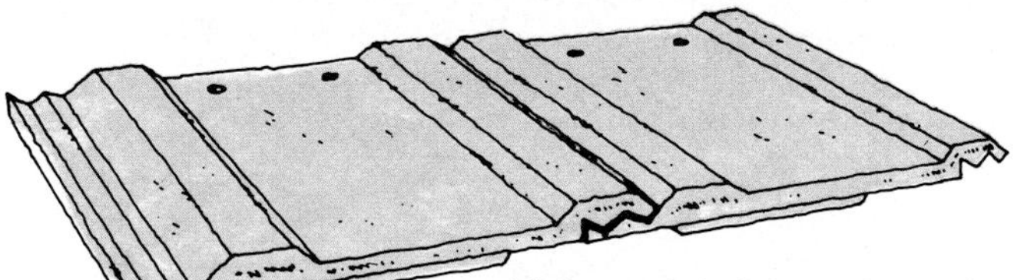

Flat concrete interlocking tile

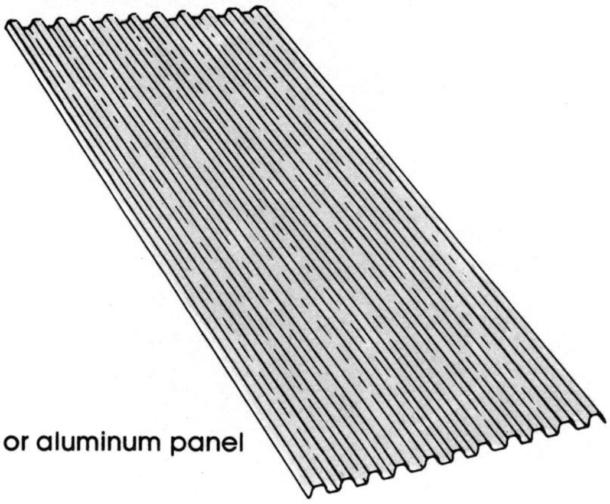

Steel or aluminum panel

are smooth-surfaced on the upper half and mineral-covered on the lower half. When the roofing is applied, the mineral-covered surface overlaps all the smooth surface of the strip below. This overlap provides excellent protection on very low-sloped roofs. Two rolls cover one square. Each roll weighs 55 to 70 pounds.

Roofing felt. This is the term for the asphalt-impregnated underlayment paper used with such roofing materials as asphalt shingles, shakes, and tile. These rolls are classified by weight per square, with the most widely used type being the 15-pound felt (that is, the material to cover one square weighs 15 pounds) and the heavier 30-pound felt. One roll covers either two squares or four, depending on the size of the roll.

Clay and Concrete Tiles

For years the curved clay tiles used on the Spanish missions of California were the standard when it came to a clay tile roof. Strong, beautiful, and heavy, they lasted as long as the house. They are being challenged now by concrete tiles that offer many advantages. Lighter, cheaper, and easier to install, concrete tiles come in a wide variety of colors. They can be barrel-shaped, like the mission tiles, flat, ribbed, or S-curved. Some varieties of concrete tile are light enough to go on a standard roof. Clay tiles, on the other hand, weigh about 1000 pounds per square. If your house wasn't originally built to handle a clay tile roof, it would have to be reinforced to withstand the weight.

Tiles are more expensive than asphalt shingles and more difficult to apply, but they never wear out.

Aluminum and Corrugated Steel Panels

These types of roofs are long-lasting and relatively easy to apply. Corrugated steel roofs have been made for more than 100 years. Although quite inexpensive, they are heavy and will eventually rust. Aluminum panels are replacing the corrugated steel roofs because they are strong, will not rust, and are easy to transport and cut. Both these roofs are noisy in a rainstorm and downright ear-shattering in a hailstorm.

Metal panels are efficient conductors of heat, so they must be well insulated underneath. Steel conducts heat about 1800 times faster than wood, and aluminum conducts heat four times faster than steel. Temperature changes cause aluminum to expand and contract considerably, which can mean that roofing nails will eventually work loose unless they have been properly and carefully placed.

Tar and Gravel

Used primarily on roofs ranging from flat to a 3 in 12 pitch, this kind of roof, often called a hot-mopped or built-up roof, is made of alternating layers of roofing felt and hot tar, with a protective final coat of fine gravel.

When a tar and gravel roof becomes worn, usually in less than 15 years, another one can be put directly over it, to a maximum of three roofs.

If your house is in need of a built-up roof, the job should be done by professionals. The tar is heated and must be kept at a certain temperature: too cold and it won't spread; too hot and the tar pot can explode.

CHOOSING THE RIGHT ROOF

CONTINUED

Cold-mop

Cold-mop roofing is an inexpensive—and much less effective—method of applying a built-up roof. The material is a liquid with an asphalt base. It does not have to be heated because it lacks the clays and other hardening agents that are part of a hot tar roof. This type of roofing is little used today except for repair work; standard roofing materials such as asphalt shingles, shakes, or wood shingles, which can be applied by the homeowner, are better choices.

Slate

Slate roofs are expensive and heavy, but will last a lifetime or more. If your roof wasn't originally designed to handle the weight of a slate roof, you will have to call a contractor or architect to reinforce your roof support structure. Slate makes a beautiful roof and is fireproof, but because of its weight it should be professionally installed.

Aluminum Shingles

Aluminum shingles are increasingly popular because of their fire resistant qualities and long life. They are expensive—comparable in price to wood shakes—but will last the life of the house. They may be dented by hail. Aluminum shingles are difficult to apply correctly, and you should have them installed by a professional.

Other Considerations

Before you select the material for your new roof, there are several things to consider.

Appearance, Quality, and Cost

The better the material, the more expensive it will be, but the longer your roof will last. Slate, tile, shake, and wood shingle roofs are all beautiful and of high quality, but they are expensive. However, you can reduce the price of a shake or wood shingle roof by applying it yourself, saving considerably on labor costs.

Roll-roofing, metal panels, and tar and gravel are less expensive materials, but you give up something in appearance when you use them on a residence.

Between the two extremes are asphalt shingles—attractive, long-lasting, and relatively inexpensive. These qualities, plus the fact that a do-it-yourselfer can apply them, have made asphalt shingles the most widely used roofing material in this country.

Fire Resistance

Asphalt shingles. In asphalt shingles, look for a class A rating. Underwriters Laboratories (UL) rates shingles A, B, or C, with A being the highest in fire resistance. Asphalt roofing manufacturers voluntarily send samples of their products to the independent UL firm to be tested. UL ratings are as follows:

- ☐ Class A: capable of withstanding severe exposure to fire
- ☐ Class B: capable of withstanding moderate exposure to fire
- ☐ Class C: capable of withstanding light exposure to fire

In order to qualify for a UL rating of any class, the shingles being tested by flame must not blow off or fall off the roof as flaming brands, must not break or crack to expose the roof deck, and must not allow continued flaming on the underside of the roof deck.

Roofing manufacturers that are granted a UL rating are proud of it and will prominently display it on each bundle of asphalt shingles. Don't use asphalt shingles that are not so marked.

Wood shingles and shakes. Unless treated with a retardant, shakes and wood shingles are quite susceptible to fire. The UL-approved, fire-retardant, pressure-treated kind are more expensive but are safer in fire-risk areas. They may be required by local building codes.

Slate, tile, and metal panels. These materials are fireproof, not just fire resistant. However, metal has a low fire rating because it conducts heat so readily.

Tar and gravel and roll-roofing. Both materials are fire resistant. Prolonged exposure to fire, however, would eventually cause either to burn.

Roofing materials

Wind Resistance

Asphalt shingles. Underwriters Laboratories also tests asphalt shingles for wind resistance. The shingles are subjected to a 60-mile-per-hour wind for two hours. To qualify for a UL listing in wind resistance, not a single tab must lift during the two hours.

If you live in a windy area, look for such a UL label on the wrapping of each bundle.

Wood shingles and shakes. Their rigid form makes these materials resistant to wind, but over time, they will work loose from their nails. How long this takes depends on how well they were nailed on in the first place.

Roll-roofing. Roll-roofing is susceptible to being pulled loose by wind, so particular care should be taken to nail and cement it securely when it is installed.

Slate and tile. Due to their weight, slate and tile are very wind resistant.

Metal and vinyl panels. If a corner of a panel works loose, the whole thing will rip off in strong wind. It's important, therefore, to install the panels securely; in an area where high wind is common, it would be wise to use longer than ordinary nails.

Tar and gravel. Built-up roofs are very wind resistant.

Roof Construction

The strength of your roof rafters is another consideration when planning a new roof. If your roof is now covered with asphalt or wood shingles, but you want clay tiles or slate, it is unlikely that the rafters can support the load. You could switch to the lighter concrete tiles, or the problem might be remedied with a support system in the attic. A contractor can advise you on this.

Another consideration is the slope of your roof. A roof that doesn't shed water rapidly can spring leaks. The lower the pitch of your roof, the more precautions you must take to avoid leaks.

Asphalt shingles are commonly used on roofs with slopes of 4 in 12 or more. On roofs with pitches of less than 4 in 12, wind can force rain under asphalt shingles, or tear them away altogether.

You'll also need to take special precautions when applying asphalt shingles on very steep slopes, over 21 in 12, as detailed on page 43.

Use roll-roofing on slopes of 1 in 12 or 2 in 12. See pages 52–53 for application instructions.

Wood shakes and shingles and tile should not be put on slopes of less than 4 in 12 since they are too vulnerable to wind-driven rain at the lower slopes.

Metal panels can be put on slopes as low as 2 in 12; on low-slope roofs, they should overlay by at least 18 inches to prevent leaks.

Color and Pattern

You can apply shingle in a number of patterns. Some result from the shape of the shingle itself; investigate the shingles available at local suppliers to see what choices you have. Other patterns are created by the way you nail the shingles on. Three common patterns are explained on pages 38–40; ask your dealer about others. Consult the chart at right for guidance on colors.

Coordinating Colors

For a harmonious exterior, the roof's color should compliment that of the siding. The chart below, taken from information provided by the Asphalt Roofing Manufacturers Association, offers some suggestions for appropriate color combinations.

Siding Color	Roof Color
White	White, black, brown, green, gray, red, beige
Ivory	Black
Beige	Brown, green
Brown	Brown, green, beige
Yellow	White, black, gray, brown, green
Deep gold	Black
Coral pink	White, black, gray
Dull red	Gray, red, green
Light blue	Red
Gray-blue	White
Medium blue	Black, brown
Deep blue	Gray
Light green	White, gray, red, brown, green, beige
Olive green	White, black
Dark green	Gray, green, beige
Gray	White, black, gray, red, green
Charcoal	White

ESTIMATING MATERIALS

Roofing Material

To order the necessary amount of roofing material, you must calculate the square footage of your roof.

Basically, square footage is found by multiplying length times width. A shed roof that is 10 feet wide and 12 feet long is 120 square feet. To calculate a simple gable roof, just measure and calculate the square feet on both sides of the roof and then add them together.

If you have a fairly simple roof that you can readily walk around on, just take exact measurements. Many roofs, however, are not simple and it may be difficult to clamber about them taking measurements. If this is the case, roof estimations may be done from the ground. Measure the roof as if it were projected flat upon the ground, then correct for the amount of slope in the roof by using the area/rake conversion chart at right.

Ordering materials for a roof also means ordering starter strips, drip edges, valley flashing, and hip and ridge shingles or tiles. The calculations necessary to determine quantities of these materials can also be made with the conversion chart.

Calculating Roof Area

Using the house illustrated opposite as an example, first measure the outline of the house from eave to eave in order to include all the roof overhangs. Transfer these measurements to a piece of paper, then put in the ridge and all the dormers, porch extensions, and chimney.

On paper, you now have a roof outlined as if it were flat. But since it is sloped, it actually covers much more area than it would appear to on your outline. The main part of the house roof, with a 9 in 12 pitch, has two different sections, one narrower than the other. On a strictly horizontal plane, the two areas are:

26′ × 30′ = 780 square feet
19′ × 30′ = 570 square feet
Total: 1350 square feet

Next you must deduct the chimney and the triangular area of the ell roof that projects into the main roof.

chimney: 4′ × 4′ = 16 square feet
ell roof: ½ (16′ × 5′) = 40 (triangular area)
Total: 56 square feet

Deducting this from the gross of the main roof leaves 1294 square feet. The total on the 6 in 12 roof is:

20′ × 30′ = 600 square feet
½ (16′ × 5′) = 40 square feet
Total: 640 square feet

The dormer can be calculated separately; but unless you have many of them, just include it as part of the gross roof area and your calculations will be close enough. If you have any skylights, deduct that area from the total.

In this example there are 1294 square feet in the main roof and 640 in the ell roof. Using the area/rake conversion chart at right to determine the conversion factor, you then multiply the total number of square feet times the conversion factor. For the 9 in 12 roof, it is:

1294 square feet by 1.250 = 1615.5 square feet

For the 6 in 12 roof it is:

640 square feet by 1.118 = 715.5 square feet

That makes a total of 2331 square feet. Add 10 percent for waste, for a total of 2564.1 square feet. Divide that by 100 and you find you need 25.6 squares. Make it 26 squares so you will have some on hand for repairs if necessary.

Eaves and ridges. These are horizontal and can be measured directly, without using a conversion chart.

Rakes. To measure the rakes, first use the drawing to find the horizontal distance. Now multiply that times the conversion factor in the chart, depending on the pitch of that roof. Add the length of the rakes to the eave measurements to find the total drip edge needed.

Hips and valleys. Because hips and valleys involve sloped distances, another chart (at right) must be used to find their true distances. Find the horizontal distances of hips and valleys on your drawing by noting how far they extend into the roof (e.g., one-half, one-third). Then, using the overall width of the roof, make a close estimate in number of feet. Multiply that by the conversion factor related to the roof pitch.

From these figures, your roofing supplier can give you the proper lengths of valley flashing and quantities of ridge and hip shingles or tiles.

Additional Materials

Measure the size of all vent pipes and buy the metal and neoprene rubber flashing units that slip over them.

If flashing is needed around a chimney or skylight, measure the linear amount needed.

You will need roofing cement, which comes in tubes, 1-gallon, or 5-gallon quantities. For a 1500-square-foot roof you would use about 2 gallons of roofing cement. If you don't use it all, seal it tightly; it will last for years and is excellent to have on hand for emergency repair.

For nails, allow 2½ pounds per square for asphalt shingles, 2 pounds for wood shakes, shingles, or tile.

Be prepared for rain. Make sure you have some rolls of 6-mil plastic on hand to spread over the roof, and 1 by 4 boards to tack down the edges of the plastic to prevent its being ripped away by wind.

For asphalt shingles you also need what is called a starter roll. This mineral-coated asphalt roofing material comes in a roll about 8 inches wide (wider strips are available, and in heavy snow areas may be required) and is nailed down at the edge of the eaves and then shingled over. It provides needed roof protection under the first row of shingles. On wood shingles and shakes, instead of the starter roll, the first course is doubled. For roll-roofing and for metal and vinyl panels, no starter course is required.

Area/Rake Conversion Chart	
Slope (inches per foot)	Rake/Area Factor
4	1.054
5	1.083
6	1.118
7	1.157
8	1.202
9	1.250
10	1.302
11	1.356
12	1.414

Hip and Valley Conversion Chart	
Slope (inches per foot)	Hip/Valley Factor
4	1.452
5	1.474
6	1.500
7	1.524
8	1.564
9	1.600
10	1.642
11	1.684
12	1.732

How to calculate roof area

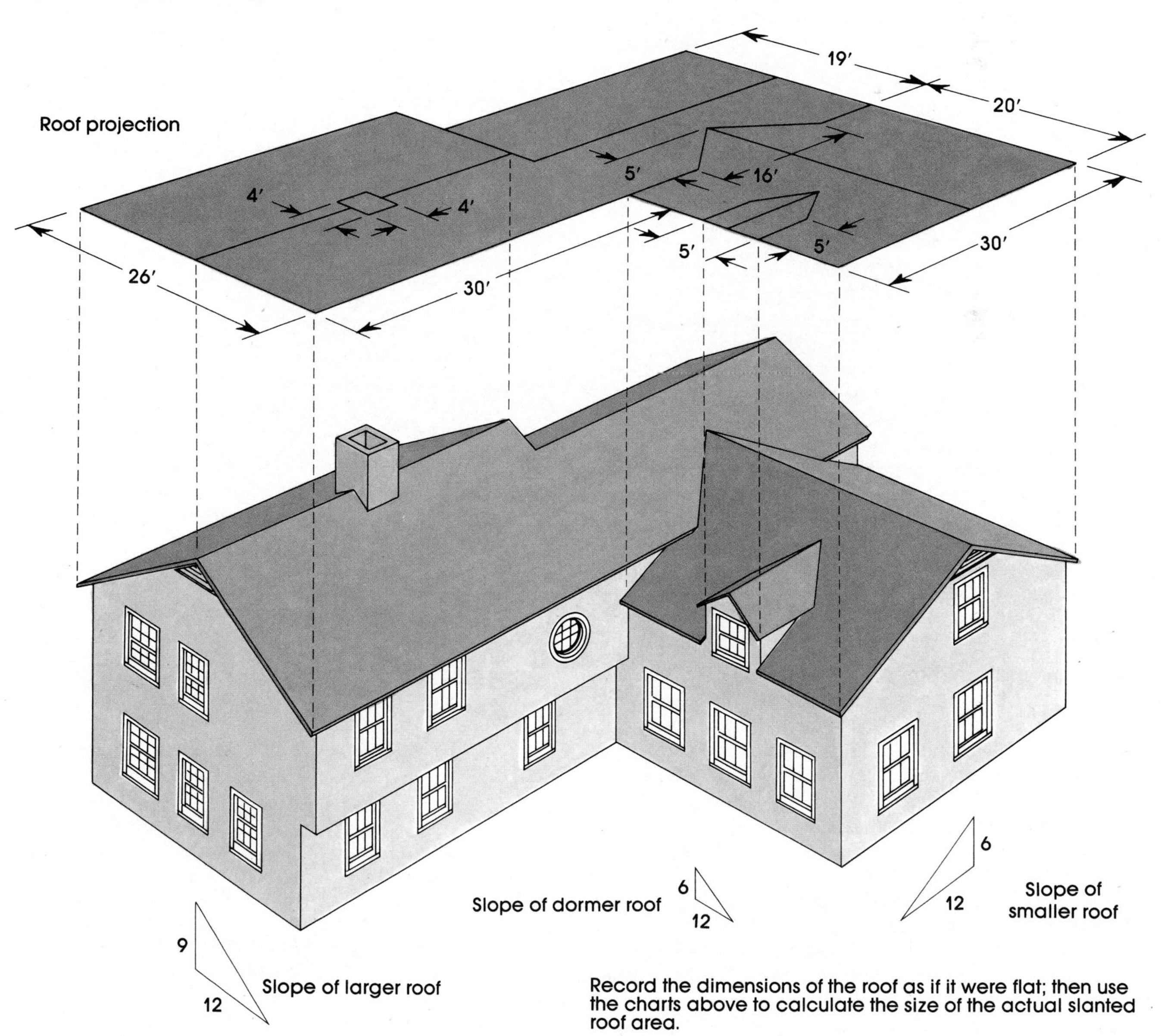

Record the dimensions of the roof as if it were flat; then use the charts above to calculate the size of the actual slanted roof area.

REPLACING AN OLD ROOF

If the roof is smooth enough that a new roof can be applied without unevenness showing, then leaving the old roof on provides additional insulation and protection and saves a lot of cleanup. But if the existing roof is badly worn and uneven, it's best to remove it. This section will deal with determining whether you can cover your present roof with new roofing, and if not, how to tear it off.

Covering an Old Roof

You can apply the new roof directly over the old if certain criteria are met.

Asphalt shingles. You can apply a new asphalt shingle roof directly over the old once you nail down the warped shingles and replace the missing ones. You can apply wood shingles over asphalt shingles if you first nail on spaced 1-by-4 boards so air can circulate under the shingles. You can apply shakes directly to an asphalt roof since their irregularities allow enough air to circulate. You can apply metal and tile—if your roof is designed to hold the weight of tile—directly over asphalt shingles. You should not attempt to put roll-roofing over asphalt shingles.

Tar and gravel. This roof is probably a 4 in 12 pitch or less, and that low slope somewhat limits your choices. Since most codes only allow three built-up roofs on a house, you should first pull off the fascia board along the rake and count the layers of exposed gravel to determine how many roofs are already in place. If there are three already, you must tear it all off before applying any new roof. In most cases, you should tear off the old built-up roof anyway—these roofs are usually so uneven that the new roof will also appear uneven.

But if you decide it is smooth enough, and it has at least a 3 in 12 pitch, you could apply asphalt shingles over it, if you have roofing nails long enough to penetrate ½ inch into the roofing deck. If it is a 4 in 12 pitch, you could apply wood shakes directly or wood shingles once you have nailed on spaced 1-by-4 sheathing boards. You could also apply a metal roof directly over it. The roof will probably not be strong enough for tile unless you remove the tar and gravel and use lightweight tile; roll-roofing should not be installed over tar and gravel.

Wood shingle. If the roof is in good condition but not even, you can improve it by nailing beveled strips of 1 by 4 or 1 by 6 against the shingle butts. These strips are called "horsefeathers" or "feathering strips." You could then cover the roof with asphalt shingles, shakes, or metal panels. Before you lay a new wood shingle roof over the old, you should first nail on the spaced 1 by 4 to provide good air circulation. Do not put roll-roofing over wood shingles.

Shake. This roof should be torn off. Shake roofs are usually too irregular to cover with anything new.

Roll-roofing. Assuming slope conditions are satisfied (see page 25), you can apply almost any roofing material over roll-roofing. Wood shingles require spaced 1-by-4 sheathing boards, and new roll-roofing should be put over an old roof only if it is still even.

Tile or slate. These materials are so long-lasting that it's unlikely you'd need to replace them. If you wish to put on a new roof for appearance's sake, remove the old one so as not to overburden the roof.

Metal or vinyl panels. Always remove panels before applying a new roof.

Roof Condition

In addition to the criteria above, you also must check that your roof is in good condition. Following the instructions on pages 12–13, inspect the roof deck for signs of deterioration, which indicates a leaking roof. Go into the attic and look for any signs of moisture or wood rot. If you see any, or if plywood laminations are separating, this must be repaired first. That means pulling off the existing roof.

Outside, check the eaves for wood rot. If you find some here but the rest of the roof appears fine, you can strip off the first two or three courses of shingles, cut out the damaged decking and replace it with matching material, and then shingle over the existing roof.

Carefully inspect the existing roof either by going up there or by examining it with a pair of binoculars. Note any loose, broken, or curled shingles. They must be made to lie flat before new roofing can go on. If the existing roof is composed of the old interlocking style of shingle, it will probably be too irregular to cover with any other roofing material.

Replace any broken and missing shingles to provide an even nailing surface. If the drip edges are rusted or broken, remove them and install new ones with the new roof. Any trim boards that are rotten or broken should be replaced.

If you plan to cover wood shingles with asphalt shingles, inspect for warping, which will make the roof uneven. Warped shingles should be split, then nailed down flat (see pages 86–87). Hammer down all protruding nails and replace any missing shingles.

Preparing to cover a wood-shingle roof

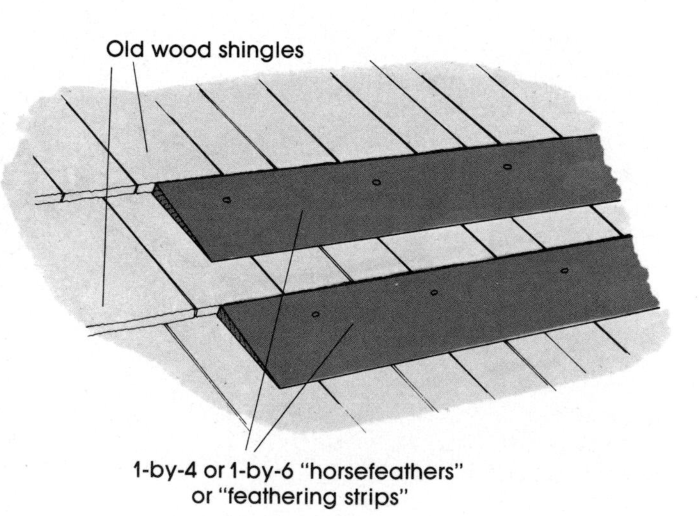

Codes and Permits

Building codes generally stipulate that a residence can have no more than three roofs on it: the original and two follow-ups. Before deciding to put another roof on the old, check how many already exist. You are talking about a lot of weight. Say you have 1800 square feet of roof to cover with asphalt shingles weighing 240 pounds per square. That's 4320 pounds on the roof. If there are already two roofs on and you add a third, that amounts to 6½ tons of roofing material. It's certainly something to think about during high winds and heavy snowfalls!

Do you need a building permit when adding a new roof or siding? In some areas the answer is no; in others, it's yes. Play it safe and call the inspector.

Removing an Old Roof

There is no special trick to removing an old roof, just a lot of hard work. This is a good time to hire a few husky teenagers in the neighborhood. Convince them you are giving them an opportunity to build their biceps.

The primary tool in removing a shingle, built-up, or roll-roofing roof is the flat-bottom shovel. Roofing outlets often sell a shovel-like tool that has serrated edges along the tip to cut nails.

Start near the ridge. On built-up roofs, use a pick to break open a line along the ridge so you can get the shovel underneath and begin prying up the material.

On wood shingle or shake roofs, be sure to work your way from the top down so debris won't fall through the open sheathing.

With shakes and shingles, a crowbar often works better than a shovel. By running the flat end of the crowbar up under the shingles, then prying up, you can remove a dozen or more shingles at a time.

Tile can be removed by hand. Slate should be pried up with a crowbar.

To remove metal or vinyl panels, use a crowbar to pry up the panels; pull the nails out.

Removing an old roof creates a large amount of debris to dispose of. The best method is to obtain a debris box. These come in various sizes and are rented by the week. Have the box placed as close to the house as possible so you can dump the material into the box directly from the roof. Having to throw it on the ground and then pick it up again is doing it the hard way. Be aware that you are responsible for any cracks in sidewalks or water pipes caused by the weight of the box or the truck that hauls it.

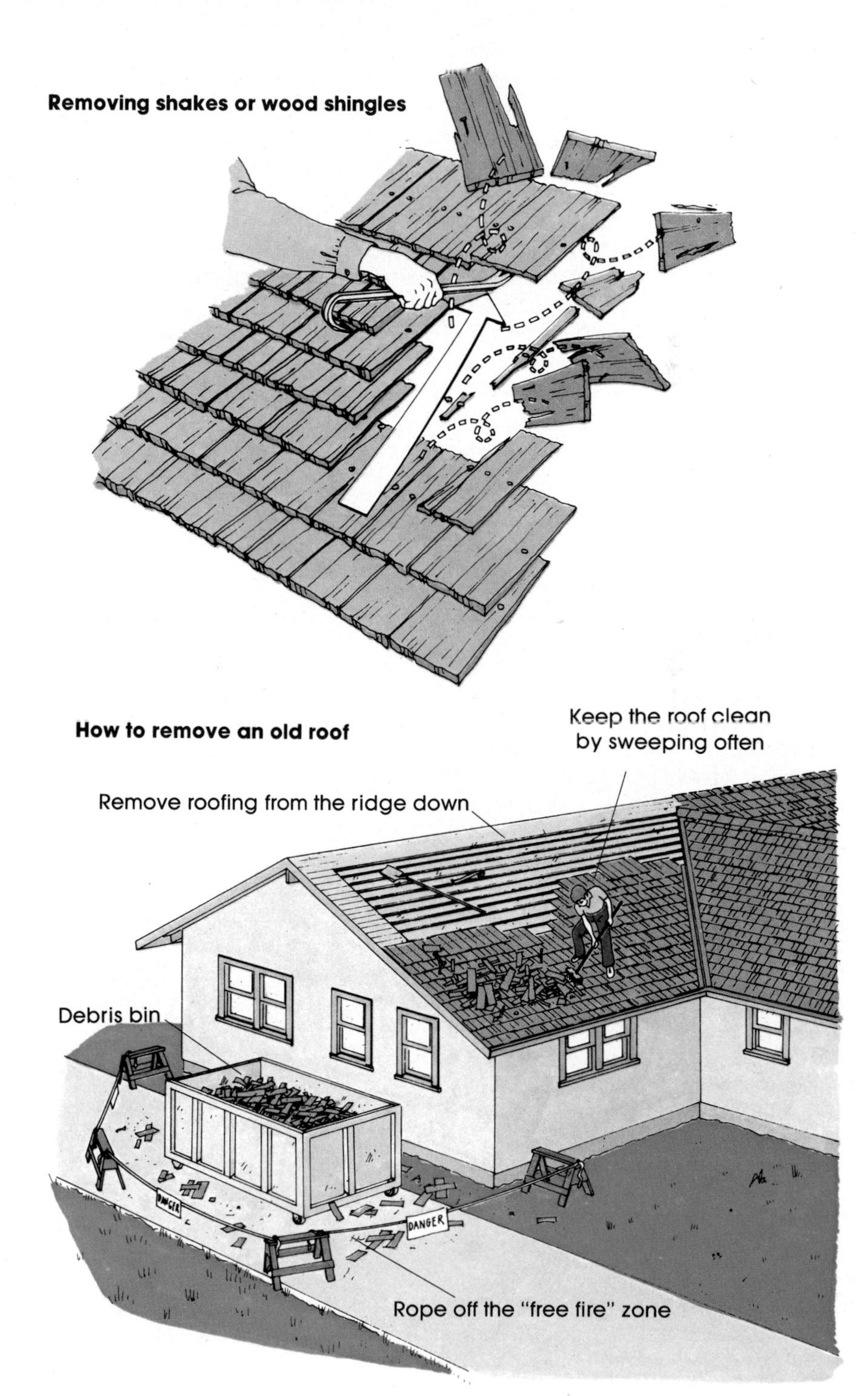

ROOFING ON NEW CONSTRUCTION

Roof Decks

There are basically two types of roof decks (or sheathing): solid or spaced. Solid sheathing is plywood or tongue-and-groove material; spaced sheathing is 1-by-4 or 1-by-6 material.

In most cases, solid sheathing is used for roofs of asphalt shingles, roll-roofing, and slate. Spaced sheathing is normally used for shake, wood shingle, tile, and metal or vinyl panel roofs.

Solid Sheathing

Starter board. Also called V-rustic, starter board is commonly used over open eaves and gable overhangs in place of plywood sheathing, which would look unattractive when viewed from underneath. Starter board is 1 by 6 and similar to shiplap. When using starter board, note that there are two sides to it. The side with the beveled edge goes down, creating the V that shows between boards when they are seen from below.

The first board along the eaves should be carefully aligned along the rafter ends. Snap a chalkline across the rafters to check that they are even. Long lengths of starter board will probably be warped, so it is necessary to push or pull them into a straight line. Each end must fall in the center of a rafter, where it will butt against the next piece.

Carry the boards up about 6 inches past the house wall. Next, cut and nail starter boards up the gable overhangs. Be sure to cut them so their ends fall in the center of the second rafter, which leaves nailing space for the plywood decking.

Plywood. Roofs are commonly covered with 4-by-8 sheets of ½-inch exterior-grade plywood. Use the cheapest grade—called CDX—since it will be covered and appearance is of no importance.

The plywood sheets run lengthwise across the roof rafters. Align the first panel carefully in the bottom left corner. Double-check that all corners are in line with rafter edges and that the leading edge falls in the center of a rafter, leaving room for the next piece to be butted up there.

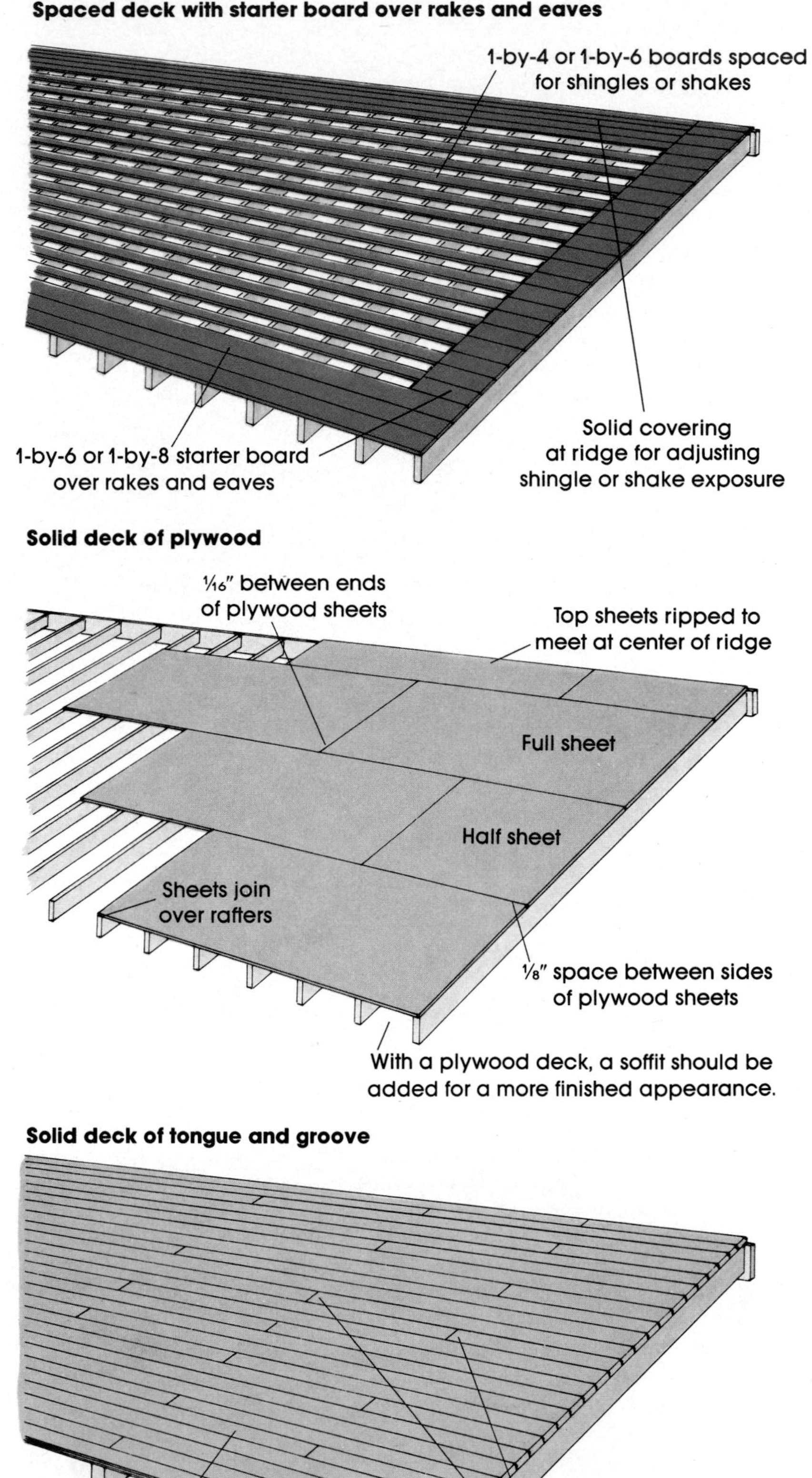

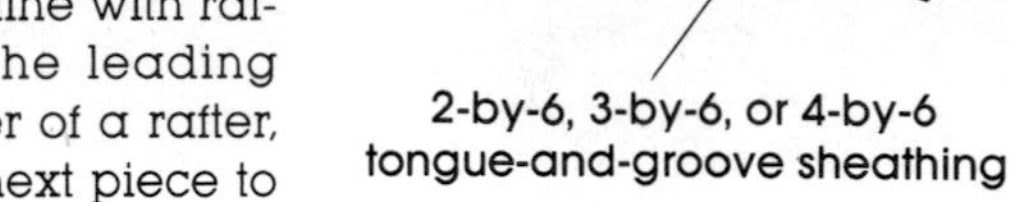

Nail the bottom course of panels in place first. Use 8d nails spaced every 6 inches along the edges and every 12 inches in the field (middle part of the panel). Plywood will expand and contract depending on its moisture content, so it's important to leave room between panels. The usual gaps are 1/16 inch between ends and 1/8 inch between sides; if your climate is very humid, double these measurements.

Start the second course with a half sheet so that all joints will be staggered. Joints that line up together on the same rafter create a weak roof.

At hips and ridges, nail the panel in place and let it hang over. Snap a chalkline directly over the hip or ridge and then cut the plywood. When you come to valleys, you will first have to measure them, then cut the plywood to fit.

If you are putting on a tile roof, check the manufacturer's instructions. In many cases, the plywood is cut back 3/4 inch on each side of the hip and ridge lines so a length of 2-by material can be placed on edge there to support the ridge or hip tiles (see illustration).

Tongue and groove. Tongue-and-groove (T&G) boards, made from 2 by 6s, are used when the roof deck will be visible from inside the house, as in cathedral ceilings. (Thus, with tongue-and-groove sheathing, starter board is not necessary.) The first board should be put down along the eaves with the tongue edge facing toward the ridge. Use a scrap piece of T&G about 2 feet long as a hammering block to knock successive boards into place. The tongue-and-groove section should fit tightly together, but leave a 1/8-inch space at all butt joints for expansion. Put two 16d nails in each board over each rafter. Stagger all joints so they don't all line up on the same rafter.

Spaced Sheathing

Spaced sheathing, which generally uses 1 by 4s, is most commonly put under wood shingle, shake, and metal panel roofs. On aluminum and corrugated steel panel roofs, the 1 by 4s are spaced every 4 feet or according to manufacturer's instructions. On wood shingle roofs, the sheathing is spaced the distance of the shingle exposure, *not* the width of a 1 by 4. Spacing is measured from center to center (abbreviated o.c.). Thus, with a 5-inch shingle exposure, space the boards every 5 inches o.c. Because 1 by 4s are actually 3½ inches wide, the spacing between boards will be 1½ inches.

On some types of tile roofs, 1 by 6 spaced sheathing is used. The spacing is dictated by the distance from the ridge to the eaves. The manufacturer provides a chart giving exact specifications.

Where appearance is important, either use starter board for the eaves and rakes, or cover them underneath with a soffit. If eaves are not covered with starter board, put the first three 1 by 4s tightly together to make an ample nailing base for the roofing material.

On shake and shingle roofs, cover solidly about 18 inches on each side of the ridge so you can make adjustments in the shingle exposure. Use two 8d nails in each board over each rafter.

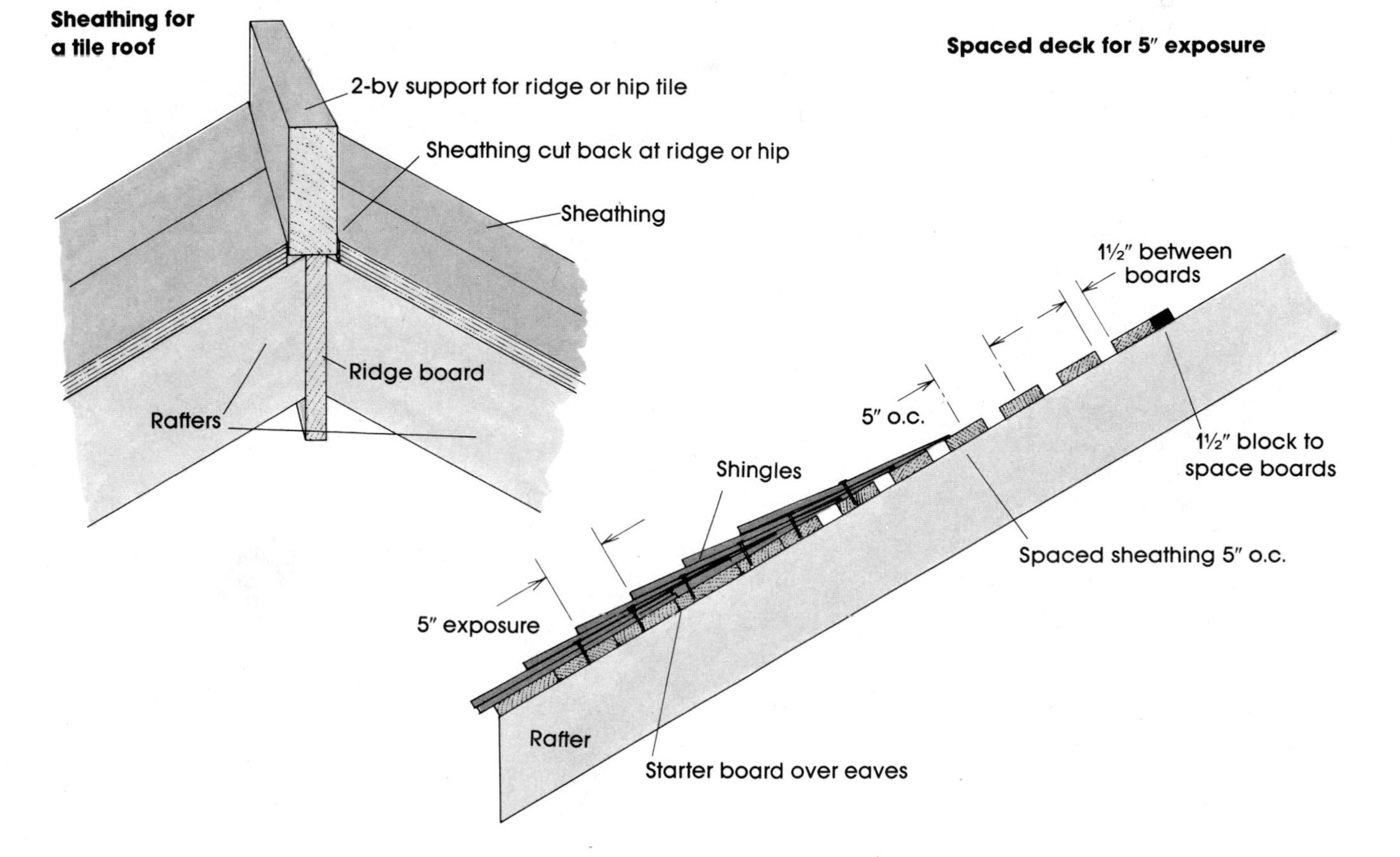

LOADING A ROOF

When preparing to put on a new roof of asphalt shingles, wood shingles, shakes, or tiles, the material must first be stacked on the roof. This is termed "loading the roof." Tiles are stacked around the roof in piles of six and spaced one tile width apart. Wood shingle and shake bundles are light enough that you can carry them up to the roof and stack them as described below. Roll-roofing can also be carried to the roof. Loading a roof with heavy asphalt shingles, however, presents more of a problem.

The best way to load a roof with asphalt shingles is to have it done by the company that sold you the material. Roofing firms have a special truck for this purpose. They just back it up beside the house, raise the truck bed on hydraulic scissors, run out a steel gangplank, and carry the bundles onto the roof. There is an extra charge for this, naturally, but it's well worth it.

If that plan isn't feasible (perhaps the truck can't get close enough to your house) and you have to load the roof yourself, you will have to rent some special equipment. Don't even think about carrying a bundle of asphalt shingles—which can weigh 70 pounds or more—up a ladder to your roof. It's unsafe and unhealthy.

What you need is a ladder loader, which usually can be found in rental shops. This is an aluminum extension ladder with a series of ropes and pulleys that raise a small platform. Do not exceed the load limit on the ladder, which is generally about 200 pounds. Two people are needed to operate it: one to load the platform and raise it, and the other to unload and stack the bundles.

If worse comes to worse and no ladder loader is available, build a ramp to the roof wide enough for two people to carry a bundle between them.

Bundles of any type of shingle or shake should be stacked on the roof so that the load is equally distributed and the roofing material is kept out of the way until needed. To do this, place one bundle lengthwise beside the ridge, and another in the same position on the other side of the ridge. Lay the next three bundles across those two, then the next three across those. Move on now, and start another pile.

Don't leave asphalt shingles out overnight or in direct sunlight for long. Shingles not being immediately used should be stored off the ground and covered with a tarp or other covering that "breathes."

Stacking bundles of shingles

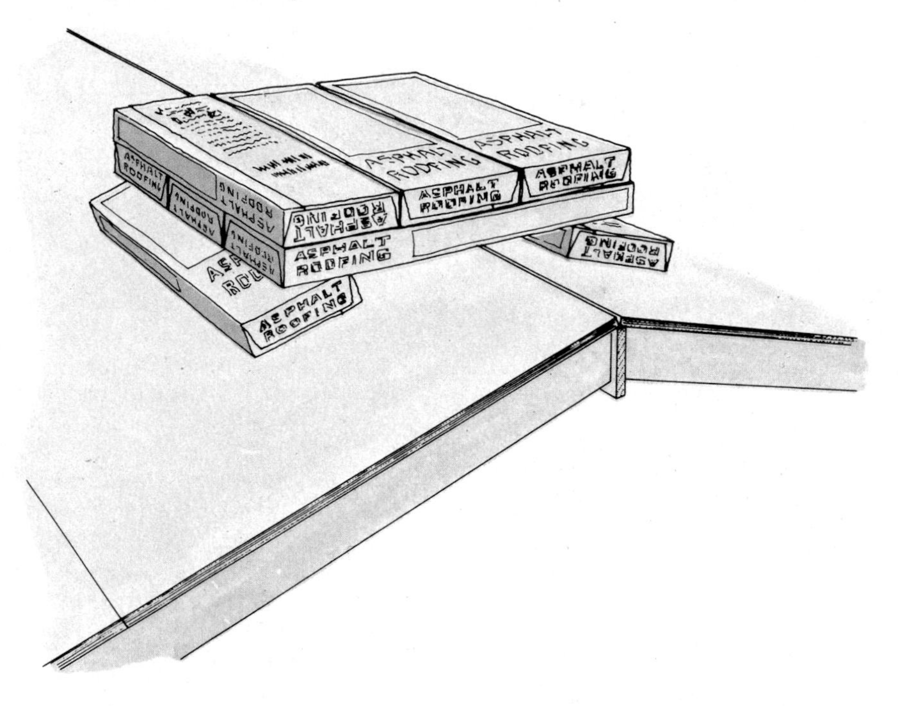

Stacking tile

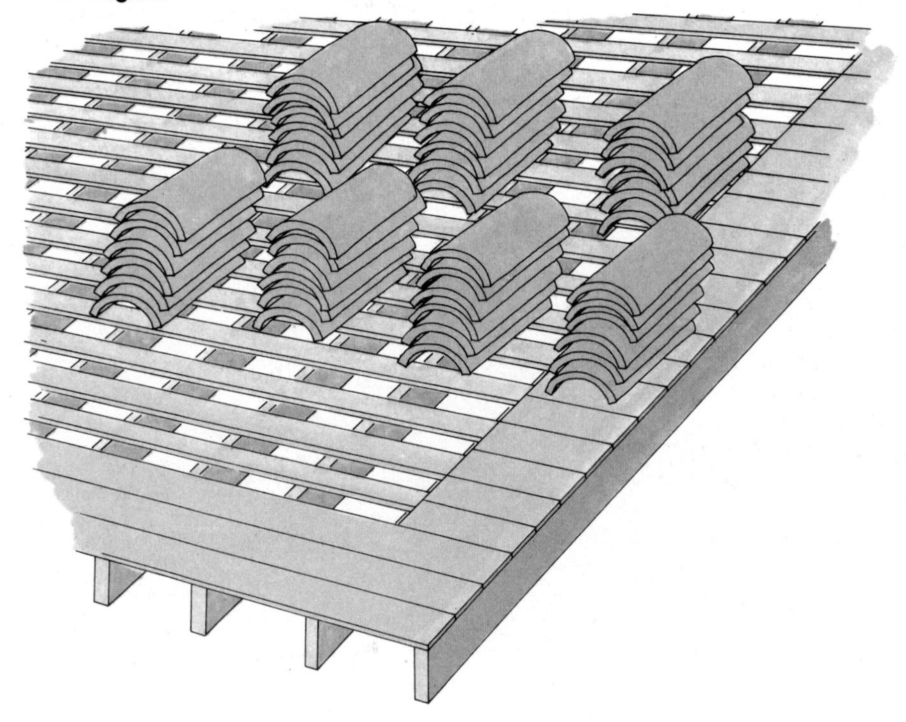

APPLYING ROOFING FELT

Roofing felt (also called underlayment) is asphalt-impregnated paper. It is named according to how much one square weighs. Roofers usually work with 15-pound and 30-pound felts.

The roofing manufacturer's instructions usually specify what type of felt underlayment is used. For asphalt shingles, one layer of 15-pound felt is best. For shakes, see page 44 for a special technique in applying 30-pound felt. Underlayment requirements for tile vary; check the manufacturer's instructions. Wood shingles, metal and vinyl panels, and roll-roofing take no underlayment.

A standard roll of 30-pound felt is 3 feet wide and 72 feet long. One roll covers two squares with a 2-inch overlap, called the headlap, on the top edges.

The felt usually has white lines printed on it. Use those 2 inches from the edge as guidelines for the overlap. Otherwise, you should generally ignore these lines. Keep shingle courses even through measuring and snapping a chalkline, as described on page 37.

It is essential that the felt be laid flat and smooth. Hot sun will soon cause it to buckle, so lay out only as much as you can cover in an hour or so. Don't put felt down over a wet deck or it will eventually bubble and distort shingles.

To lay out felt, align it on the bottom left corner of the roof and hold it in place with three roofing nails near the edge, as shown. These nails hold the paper but still allow you to straighten it.

Roll the felt to the other side of the roof, pull it smooth, and cut flush with the rake edge. Making sure it is smooth, put a nail every 3 or 4 feet along the upper half. Do the next layer the same way, overlapping 2 inches at the headlap and 4 inches on the sidelap.

When you reach the ridge, lay the paper across and tack it on the other side. At valleys and hips, carry the felt at least 12 inches to the other side and cut. Where a vertical wall meets the roof, such as around dormers, carry the felt up the vertical wall about 5 inches.

When you come to vents, roll the felt next to the pipe and note where the two meet. Cut a slit in the felt, drop the felt over the pipe, then cut the felt around the vent for a smooth fit. When you continue rolling out the felt, check that it was not pulled out of line by the vent pipe.

Here's a tip from the pros: if there are any wrinkles or bubbles in the felt that you can't pull out, make a long cut through the wrinkle and lap the top edge over the bottom. Nail the felt flat.

How to apply roofing felt

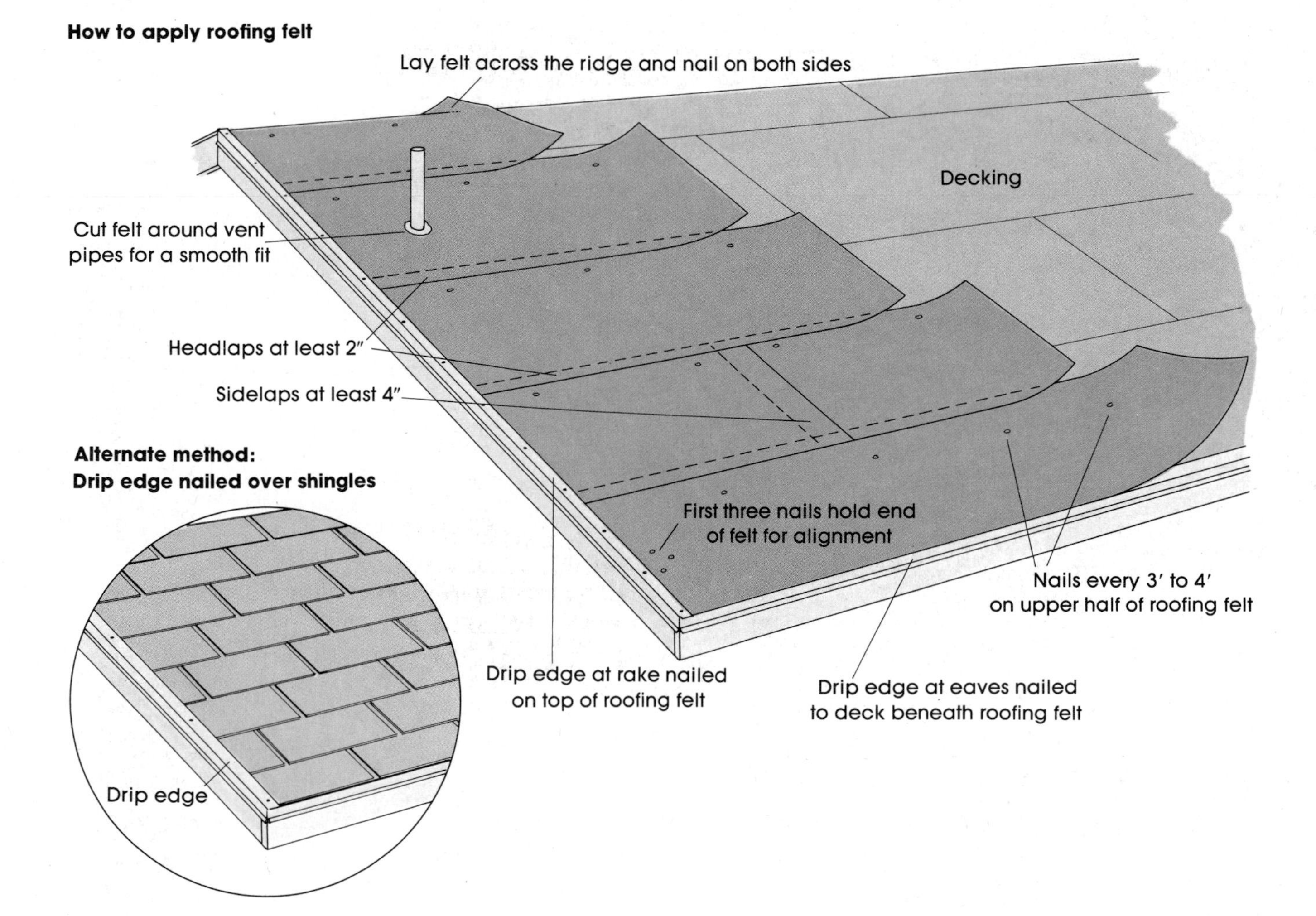

VALLEY FLASHING

There are two basic kinds of valley flashing: open, created by installing W-metal valley flashing (or, occasionally, roll-roofing), and closed, created by shingling across the valley in one of the ways described below. Which kind you should choose depends on the type of roof you have. If you have straight valleys between roofs with equal slopes, then an open valley that is made of metal flashing is fine. If the valley runs between roofs of different pitches, a full-lace or half-lace closed valley will do a better job.

Metal flashing is commonly used with wood and asphalt shingles, shakes, or tile. It is called W-metal flashing because it is shaped somewhat like the letter W. Made from aluminum or, less commonly, galvanized tin, it comes in 10-foot-long sheets that range in width from 16 inches to 24 inches. The wide type is used on lower-pitched roofs, particularly those subjected to heavy rainfall. The narrower flashing is used on more steeply pitched roofs in areas that receive moderate to heavy precipitation.

Generally, closed valleys are shingled across only with asphalt shingles, either fully laced or half-laced.

Flashing is needed in the valleys between a dormer roof and a main roof, just as in the main valleys. For asphalt shingle roofs, use the full-lace or half-lace style, as described below. For wood shingle, shake, panel, or tile roofs, use W-metal flashing. For roll-roofing roofs, use a roll-roofing open valley.

Metal open valley. The valley tin goes on after the roof and valley have been covered with roofing felt (see page 33) and underlayment, if necessary. Lay the flashing in place, with the ridge centered up the valley. Use a pair of tin snips to trim the bottom edge even with the edge of the roof along the eaves.

Nail the flashing to the roof deck, placing the nails ½ inch in from the edge of the flashing, every 6 inches. Use aluminum nails with aluminum flashing.

If the valley is longer than one length of valley tin, start from the bottom and lap the second length over the lower one by 6 inches. Use tin snips to trim the top flush with the top of the ridge.

In shingling along W-metal valley flashing, first snap chalklines as trimming guides. At the top, mark 3 inches out from the center ridge on each side. Work down, adding ⅛ inch in width for every foot of valley length. Thus, an 8-foot-long valley would be 1 inch wider at the bottom than at the top. Calculate the width at the bottom of the valley, then snap the chalklines.

As each course arrives at the valley, the top corner of the last shingle must cross the chalkline. If it will fall just a little short, adjust the course farther back. Don't let a joint between shingles fall on the flashing.

When you have shingled to the ridge, snap a chalkline on the shingles over the guideline; cut the shingles along the line. Before cementing the shingles over the flashing (do not use nails in the valley), lift each corner and trim off each sharp end. See box opposite for this important step.

Roll-roofing open valley. Sometimes 90-pound mineralized roll-roofing is used to flash an open valley. It is a little less expensive than the other methods, but it may not last as long as the roof itself.

To use roll-roofing for valley flashing, first cut a length of 18-inch-wide roll-roofing the length of the valley. Center it, mineral surface down, in the valley, and nail it one side at a time. Then lay a 36-inch-wide strip of roll-roofing over it, mineral surface up. Snap a chalkline down its center and position the chalkline over the valley crease. Nail one side first, then bend the strip to seat it securely in the valley, and nail the other side.

Snap chalklines 3 inches out

Installing a metal open valley

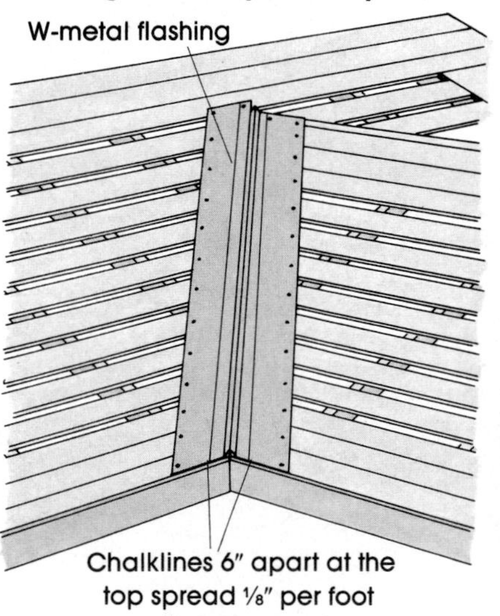

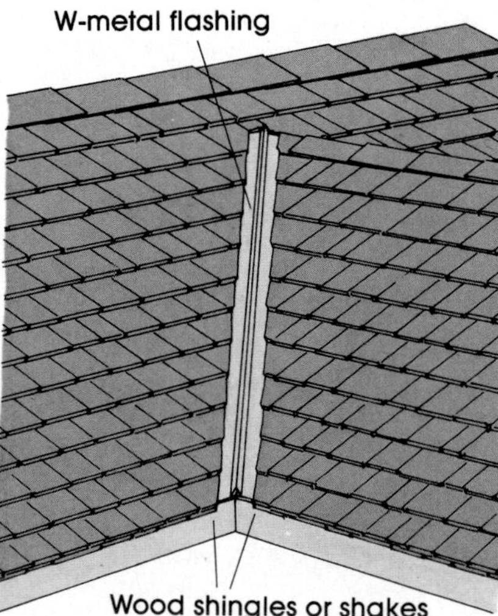

Installing a roll-roofing open valley

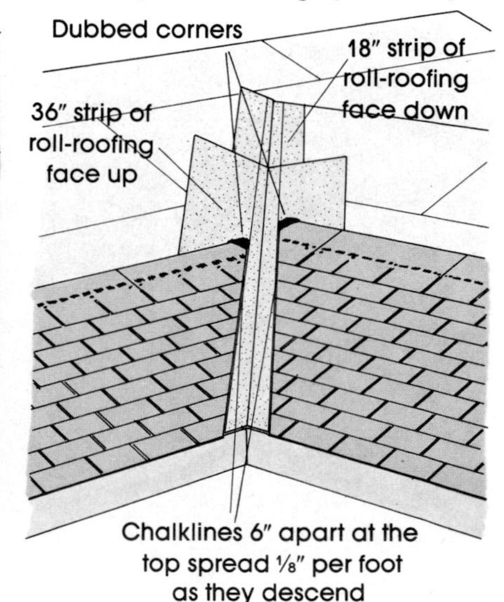

from the center on each side, diverging ⅛ inch for each foot of descent. Cut the roofing material on these lines as it overlaps the flashing. Cement each edge over the valley flashing and trim the corners, as described below.

Half-lace closed valley. A half-lace valley may not appear as finished as an open W-metal valley or a full-lace valley, but it can be installed more quickly. It works particularly well where two roofs of different pitch are joined.

Carry the shingles from the lower-pitched roof across the valley and at least 12 inches up the other side. Make sure there are no joints within 10 inches of the center on either side. If necessary, insert one tab between shingles farther back on the course to ensure that the last shingle reaches far enough up the other side.

Now bring the shingles of the steeper roof across to the center of the valley. Snap a chalkline 2 inches back from the center of the valley and trim the shingles along the line with tin snips. Trim all corners (see box below).

Full-lace closed valley. The full-lace valley is somewhat more time-consuming than the half-lace because you must shingle from both sides of the valley at once. On the other hand, you don't have to spend time trimming shingles.

If the roof slopes on both sides of the valley are equal, the full-lace valley is essentially a matter of criss-crossing one course under another as you work on the roof. As with the half-lace valley, carry the last shingle at least 12 inches up the other side, adding a single tab farther back along the course if necessary, since no joints should be within 10 inches of the valley.

If the roof slopes are different, as in the illustration, two courses of shingles on the flatter roof will cross to the other side before you bring one across from the steep slope. In some cases it may be three to one. You will be able to tell which is necessary as you work.

Installing a half-lace closed valley

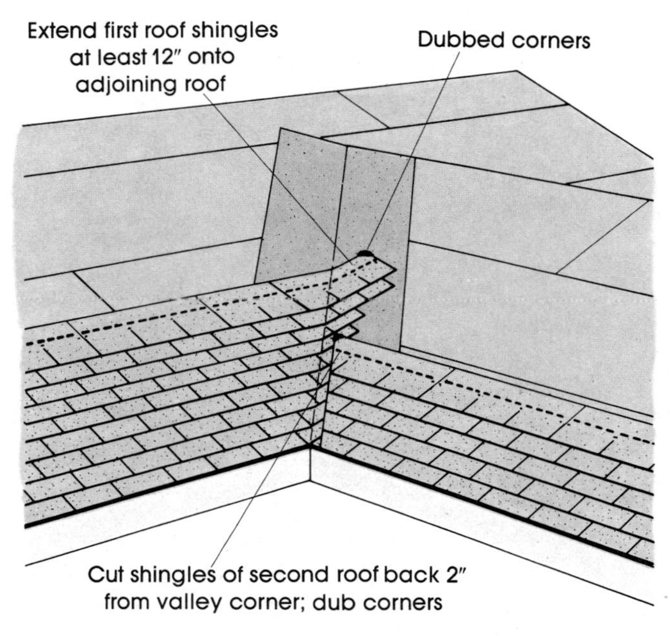

Installing a full-lace closed valley

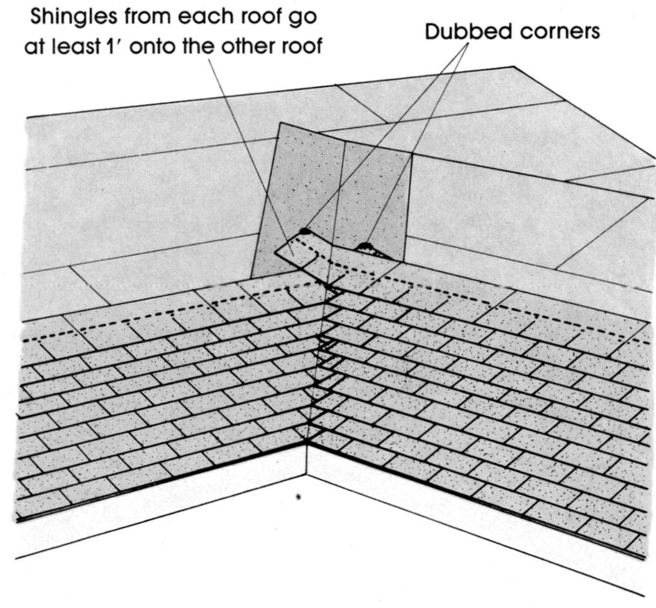

Trim Sharp Corners!

Leaks in valleys are one of the more common causes of leaks in a house. Such leaks sometimes result from the failure to trim—or "dub"—the sharp corners of shingles where they extend over the valleys.

These sharp corners can act as diverters during heavy runoff and send a course of water great distances across the roof until it begins dripping inside the house.

To dub the corners, lift each shingle and trim back the corner of the one underneath about 2 inches. Now raise and embed the edge of the underneath shingle in a 3-inch-wide bed of roofing cement.

Dubbing corners

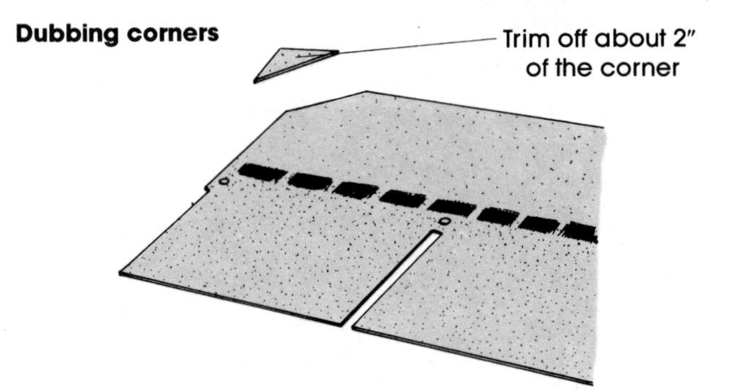

SHINGLING TECHNIQUES

Asphalt-shingle nailing pattern

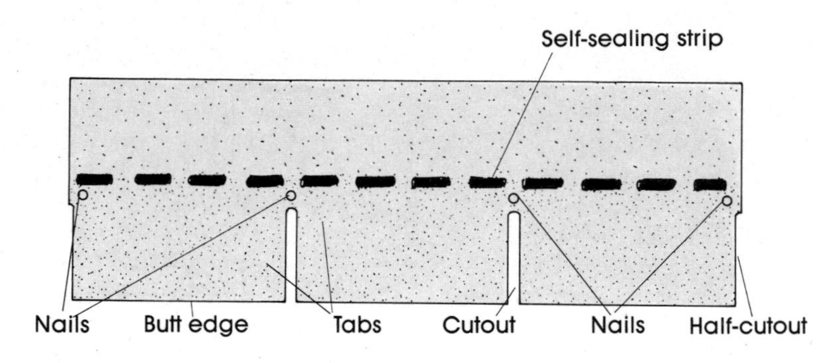

Wood-shingle and shake nailing pattern

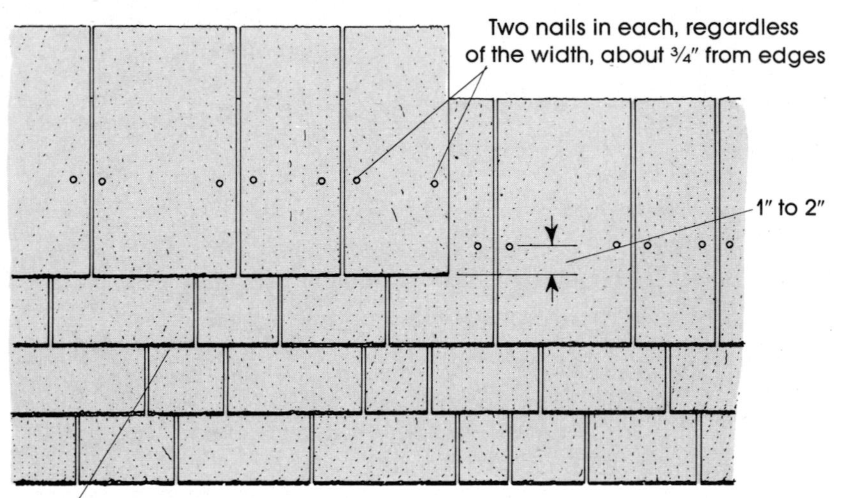

Using the gauge on a roofing hatchet

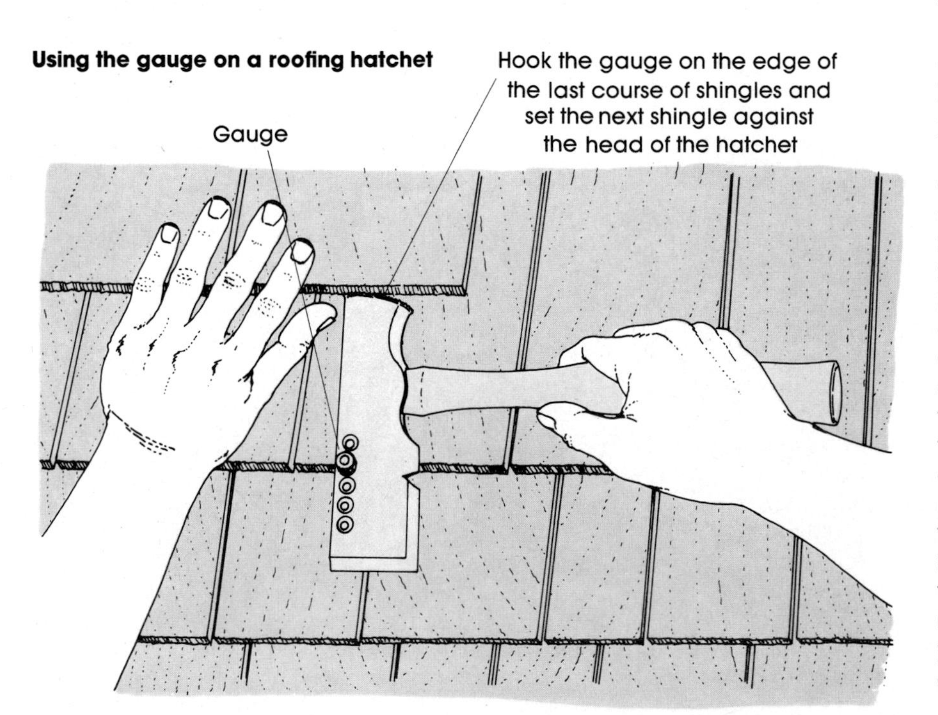

Here are some tips to use when you're putting on asphalt shingles, wood shingles, and shakes.

First, re-read the safety guidelines on pages 14–15. Be sure to wear loose pants that won't bind when you sit, and sneakers or soft-soled shoes for both a sure grip on the roof and protection against deforming the shingles, which can be a problem in temperatures over 80 degrees F. Shingling an asphalt roof is hard on the seat of your pants. For good protection, some professional roofers cut a length of inner tube long enough to reach from waist to knee and slip it around the leg they sit on.

Nailing Shingles

If you are right-handed, always start shingling from the lower left corner of the roof. Curl your left leg under you and use your right leg to maintain your position on the roof. As you shingle, you will work up and out to the right. If you're left-handed, you'll probably want to reverse directions.

By far, the most widely used asphalt shingle is the three-tab shingle. Each end of the shingle has half a cutout that, when it is fitted against another shingle, forms a full cutout.

Each three-tab shingle is fastened with four roofing nails—one 1 inch in from each end and one above each cutout. If the shingle has a self-sealing strip, place nails just below it, not in it or above it.

Wood shingles and shakes are always nailed down with two nails only, regardless of how wide they may be. Place the nails about ¾ inch from the edges and 1 inch to 2 inches above the exposure (the distance from the butt of the shingle to the butt of the shingle above). The exposure, also called weather, varies, depending on what type of shingle or shake you are using and the pitch of your roof (see page 21). Set the exposure with the help of the gauge on the roofer's hatchet. Always check the weather at both ends of the shingle—especially with the wide asphalt shingles—to keep your courses straight. Always nail in the same direction—from left to right if you're right-handed, right to left if you're left-handed. If you see that a shingle or a shake is crooked, don't try to realign it after two nails are in it. It will only buckle or split. Take it out and do it right.

It is important that the first course of shingles be perfectly straight. To align asphalt shingles, let the first one overhang the drip edge by 1/4 inch to 3/8 inch and nail it down. Place another shingle the same way at the other end of the roof. Now snap a chalkline between the two along the top edges and line up the top edges of intervening shingles with that line.

For wood shingles and shakes, place one at each end of the roof as described above, then drive a tack into the butt of each one and stretch string between them. Align the butts of the intervening shingles or shakes on the string.

With all types of shingles or shakes, check for straightness every three or four courses by measuring up from the butts of the shingles at each end of the roof and snapping a new chalkline to place the butts on for the next course.

The slowest part of roofing is simply driving all those nails. This process can be markedly speeded up by using either a nail stripper, or—for wood shingles or shakes—a pneumatic staple gun.

A nail stripper is a metal box that is strapped to your chest. The box is filled with nails, then shaken so that the nail shanks fall through slots in the bottom. Grasp a row of nails between your index and middle fingers (not the thumb and index finger), slide them out the slot on the side, and position them one at a time over the shingle. Use one tap to set the nail and one more to drive it home.

For asphalt shingles, wood shingles, and shakes, the pneumatic staple gun is even faster. The guns can be rented. Load the gun with staples, then align the shingle, press the end of the gun against it in the proper nailing spot, and pull the trigger. Be sure to hold the gun vertical to the roof so the staple is not driven at an angle.

Cutting Shingles

Shingles—asphalt or wood—and shakes must be cut in order to fit properly around vents, along the rake, in valleys, and beside flashing. Wood shingles and shakes should be cut with a circular saw for straight lines or with a saber saw for curves. Asphalt shingles should be cut with a utility knife or with the razor blade in the hatchet, if you have that type of roofer's hatchet. Always turn the asphalt shingle over and cut on the back side; cutting on the mineral surface will quickly dull your blade. Score the shingle deeply, then bend it until it breaks. Don't use a straightedge as a guide for each cut or the work will go extremely slowly. Just eyeball it and make the cut. Practice makes perfect.

Lining up the first course of asphalt shingles

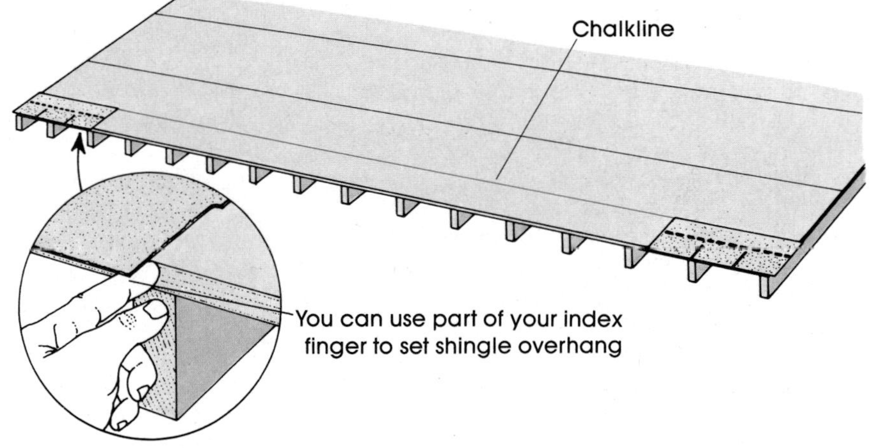

Using a nail stripper

Lining up the first course of wood shingles or shakes

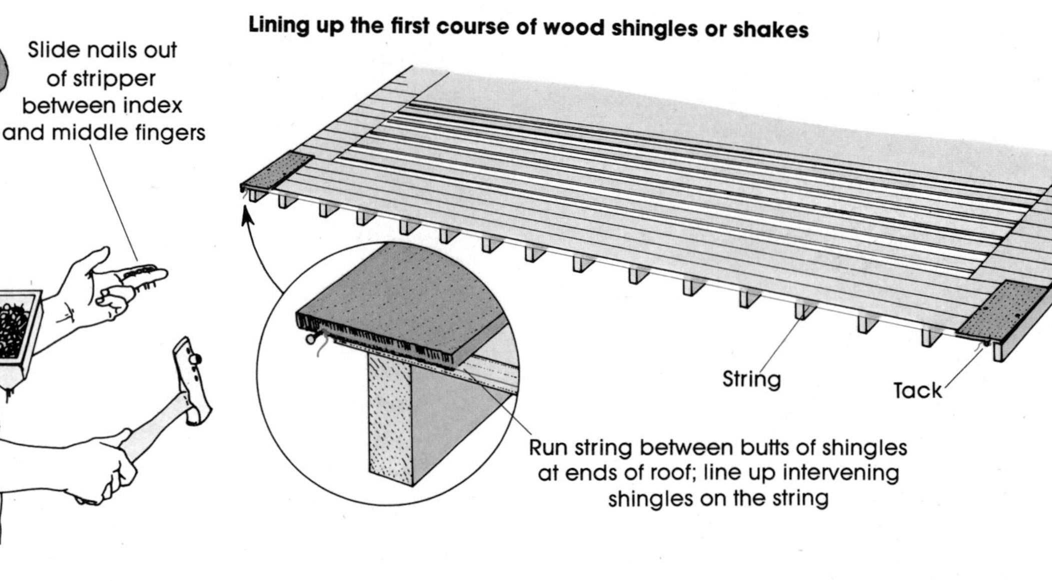

ASPHALT SHINGLES

The Art of Shingling

Applying an asphalt roof is more than simply nailing on some shingles. Making a good roof involves laying the roofing felt smoothly, applying the drip edges correctly, keeping shingle courses straight, and knowing how to flash around vents and chimneys and in valleys. All this comes with practice. Your work will start slowly—but then, you are doing it because you have more time than money. The secret to a good roof is doing careful work; mistakes mean leaks. The following directions apply to all roof types except hip roofs; see page 40 for shingling a hip roof.

Drip Edges

The first item to be put on a new roof is the drip edge along the eaves. Drip edges are L-shaped strips of aluminum or galvanized steel that are nailed along the eaves and rakes of your house to prevent wood rot in the roof deck edge. Always place the wider side of the drip edge on the roof and let the other end hang over the exposed edge of the roof deck.

Aluminum drip edging should be nailed with aluminum nails. Galvanized roofing nails will cause a chemical reaction with aluminum that will result in disintegration of both metals.

After drip edges are applied to the eaves (only the eaves, not the rakes), lay the roofing felt over the roof (see page 33).

Drip edges for the rakes may be applied in one of two different fashions. One style is to put them over the roofing felt. A second method, increasingly popular, is to put them over the shingles. This method prevents rain from blowing under the shingle edges.

Flashing

The next step is to install the valley flashing. Follow the instructions on pages 34–35; other forms of flashing will be applied later.

Starter Roll

In all types of shingling, an extra layer is always applied along the eaves. This is because shingles in subsequent courses overlap at least once for double coverage, and the first row needs a similar overlap. This layer is applied after the drip edges, roofing felt, and valley flashing are in place.

The easiest method of providing the extra layer is to apply starter roll. This is a 7- or 8-inch-wide roll of mineral-surface roll-roofing that is nailed along the eaves. It should overhang the edge of the roof by ½ inch.

If you neglected to buy the starter roll, you can use shingles as a starter course. Just reverse them so that the tabs point up the roof and the upper edge overhangs the eave by ½ inch.

Shingling Patterns

After the starter roll is nailed on, you are almost ready to start shingling. Ah, but you don't get away that easily. You now have to choose which shingling pattern you want. The three basic shingling patterns used are 6-inch, 5-inch, and 4-inch. The terms refer to the distance each shingle is offset from the one below it to prevent leaks from working through the cutouts. All patterns will protect the roof equally well.

6-inch pattern. This is the easiest style to apply and is the one shown in the directions on the bundles. At the start of each course you cut off part of the first shingle, working in 6-inch increments. The remaining shingles in each course are whole ones. The resulting pattern aligns each cutout directly above another cutout in every other course. There is one problem with the 6-inch pattern in roofs 40 feet or longer: it is difficult to keep the cutouts perfectly aligned up the roof because of inherent minor differences in individual shingles. You can keep them straight, however, by periodically snapping vertical chalklines on the roofing felt as guides, as described below.

Start the 6-inch pattern by nailing a full-length shingle in the bottom left corner. The bottom edge should overhang the eave by ¼ inch to ⅜ inch and the left edge should be flush with the edge of the rake. Put another shingle at the opposite end of the roof and snap a chalkline between the two along the top edge to make a perfectly

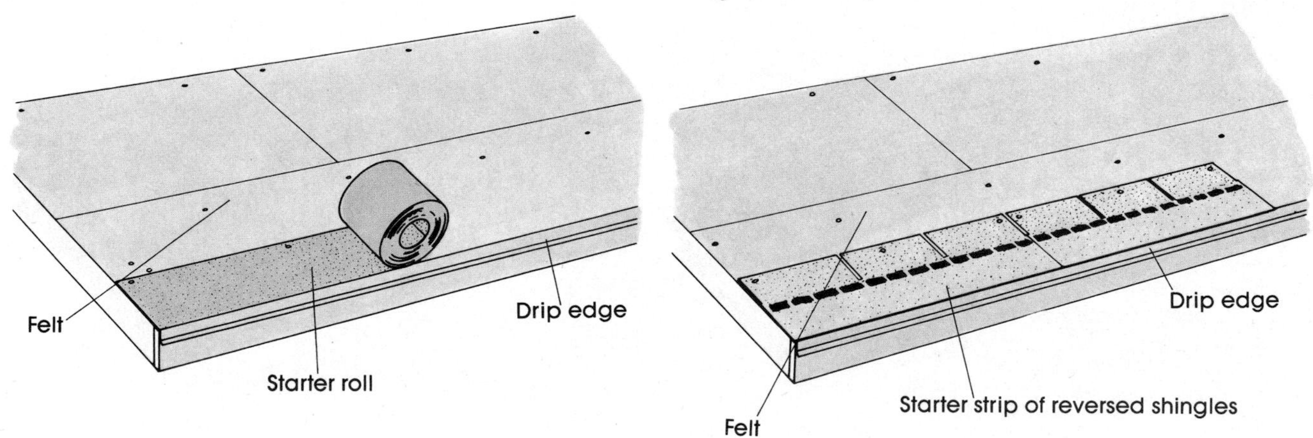

straight first course. Nail on the first course.

Now cut 6 inches, or ½ tab, off the left edge of the first shingle on the second course. Use your hatchet gauge to set the 5-inch weather (exposure) at both ends of the shingle and nail it down. For the third course, remove 12 inches, or a full tab, from the first shingle and put it in place. Continue in this manner, removing an additional 6 inches each time, through the sixth course, where you will have removed 30 inches, or 2½ tabs. Carry each course part way across the roof, far enough to keep the pattern going. Start the seventh course with a full shingle and repeat the process. When you get to the ridge, go back and fill out each course across the roof, from the bottom up.

Do not attempt to alternate whole shingles and shingles with ½ tab removed. This is a guaranteed leaky roof. When starting each course, always remove an additional 6 inches. Save the cut-off pieces and use them at the other end of the roof to finish out courses.

To keep the cutouts aligned vertically, snap two chalklines up the roof on the felt. Put the first one 36 inches in from the edge of the rake (the length of one shingle) and the second one 72 inches in (the length of two shingles). If the roof is interrupted by a dormer, chalklines must also be snapped on the other side of the dormer to keep the pattern consistent. The process of working around a dormer—"tying-in"—is explained on pages 42–43.

5-inch pattern. This pattern is widely used by professionals because the 5-inch increments are the same as the exposure. This so-called "random" pattern also eliminates the problem of aligning vertical cutouts on long roofs.

The first course begins with a full-length shingle. The second course begins with 5 inches removed from the left end. There is a trick to doing this work quickly. After the first shingle is down, put the second-course shingle on top of it, then use the hatchet gauge to move it 5 inches to the left of the

Asphalt shingle application

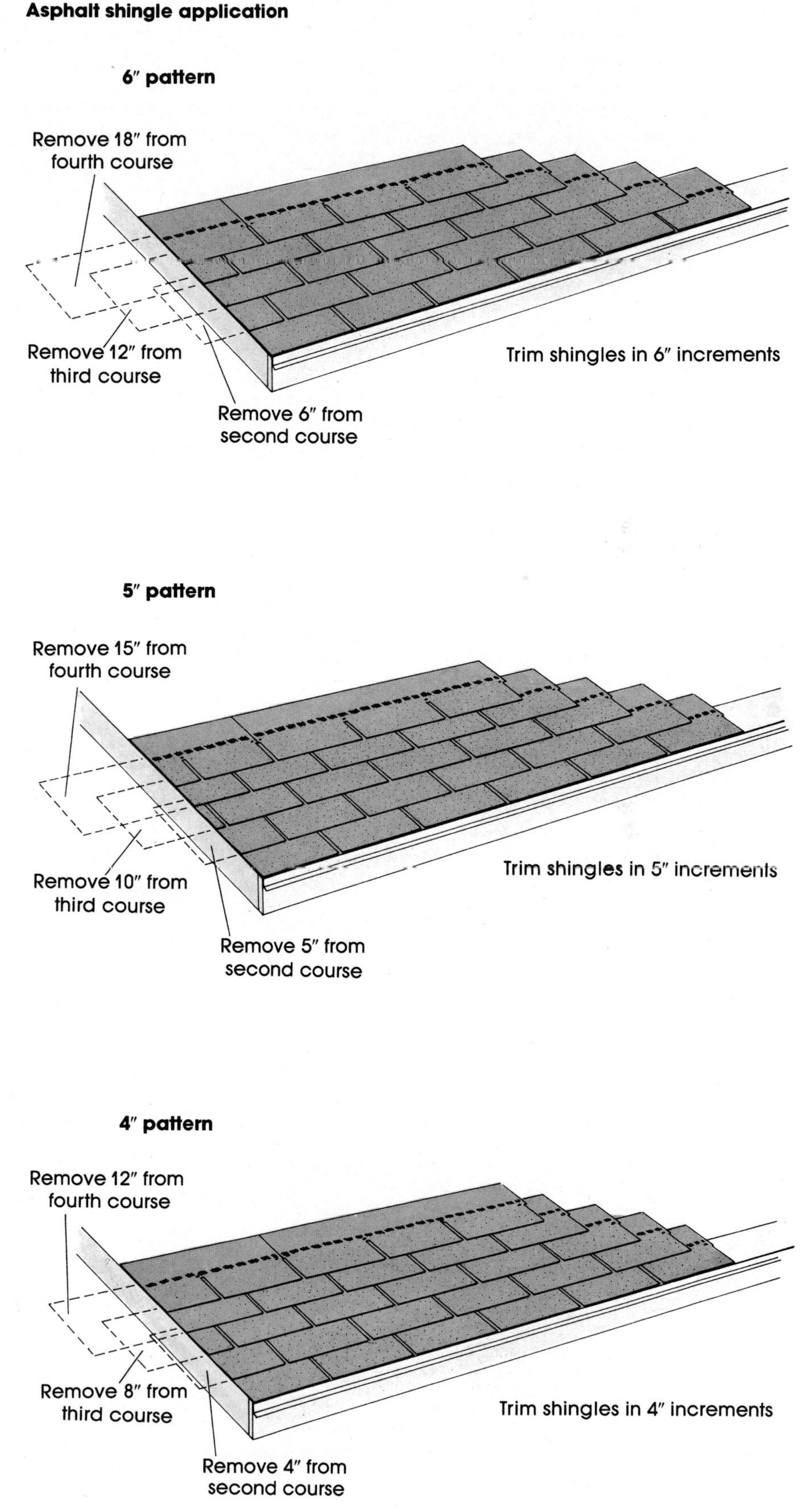

first shingle. Grasp the overhanging portion right at the rake edge, flip the shingle over, and cut it. Put the cut shingle back in place and use the gauge to adjust the exposure at each end and the distance from the end of the first shingle. Nail it down. Always set the shingle according to the hatchet gauge, not the cut end, which may be slightly inaccurate. When the roof is complete, you can trim the edge more precisely or cover it with a drip edge.

The third course is done in the same manner, offsetting it from the second course by 5 inches, which results in a total of 10 inches taken off that shingle. Continue in this manner through the seventh course, from which you remove 30 inches. Taking 5 inches off the eighth course would leave only 1 inch of shingle, which is too little to work with. Start the eighth course with a full shingle and continue.

Keep working up the roof, filling out each course only far enough to keep the pattern going. Cutting and laying shingles along the rake while you work up the roof is the slow part, but do it accurately. When you reach the ridge, go back to the eaves and start filling out each course, working from the bottom up.

Every three or four courses check that your work is not drifting out of line. Do this by measuring from the butt ends of the first course at each end of the roof up to any given course. The measurements should be the same. If one course has drifted, snap a chalkline to straighten the next course. Don't remove a crooked course unless it is radically out of line, since it will not likely be noticed.

4-inch pattern. This style is only needed on low-slope roofs with pitches such as 2 in 12 or 3 in 12. In these cases, the roofing felt should be overlapped by 19 inches instead of the standard 2 inches, for extra protection against leaks.

As with the other patterns, start the first course with a full shingle, then trim the first shingle in each successive course in 4-inch increments. Continue through the ninth course, from which you will remove 32 inches. The tenth course starts with a full shingle.

Don't try a 3-inch pattern just to be different—there should never be less than 4 inches between cutouts in any two courses.

Shingling a Hip Roof

A little more cutting is required for a hip roof. Start in the lower left corner (if you are left-handed, start in the lower right corner) and place the starter course and first course as you would on a standard roof, but along the hip cut the shingle with no overlap. Start the second course in standard fashion, cutting the shingle on the line of the hip.

Before applying the hip shingles, tack a strip of window flashing over the hip as added protection against rain.

Applying hip shingles

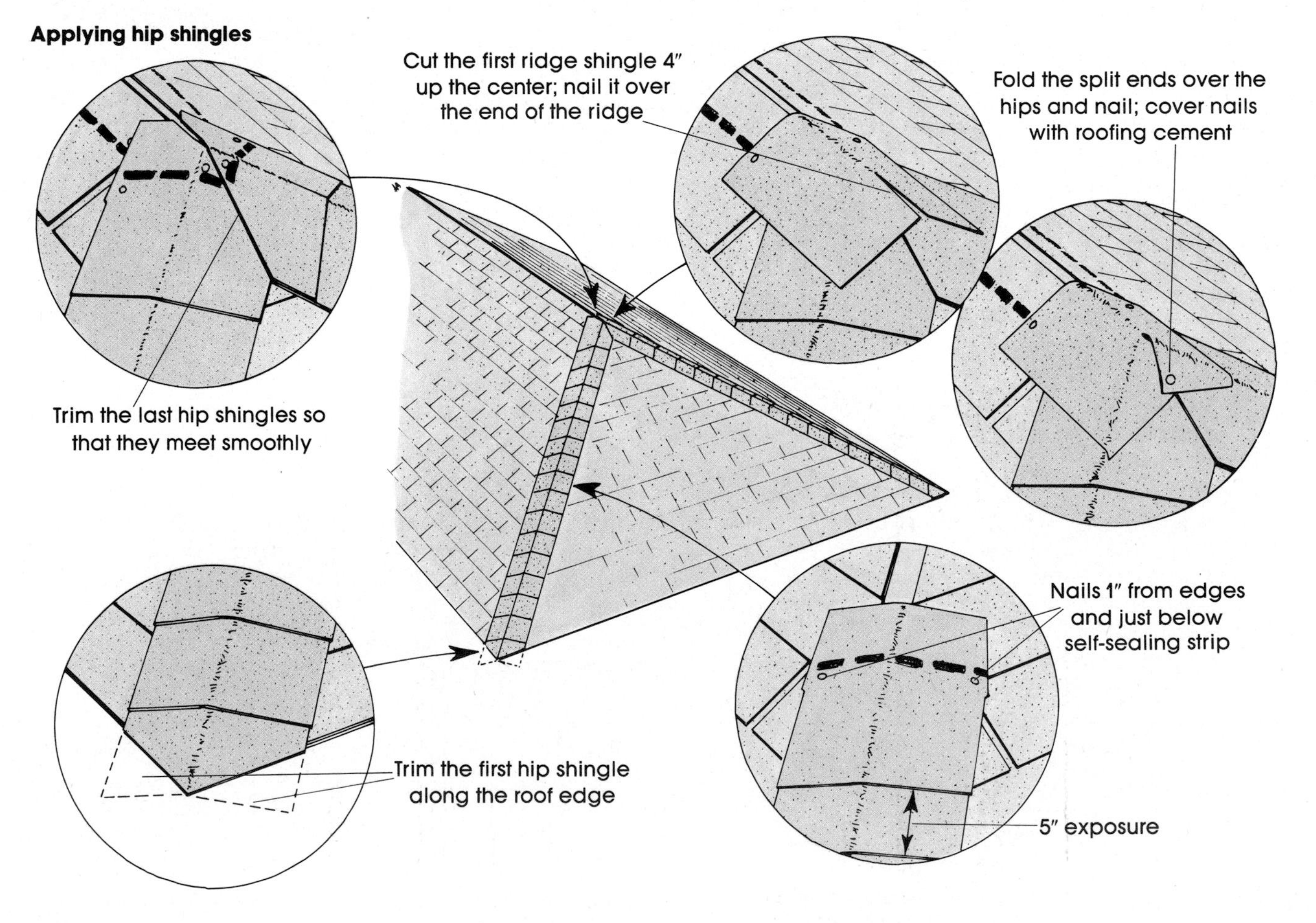

Cutting hip shingles

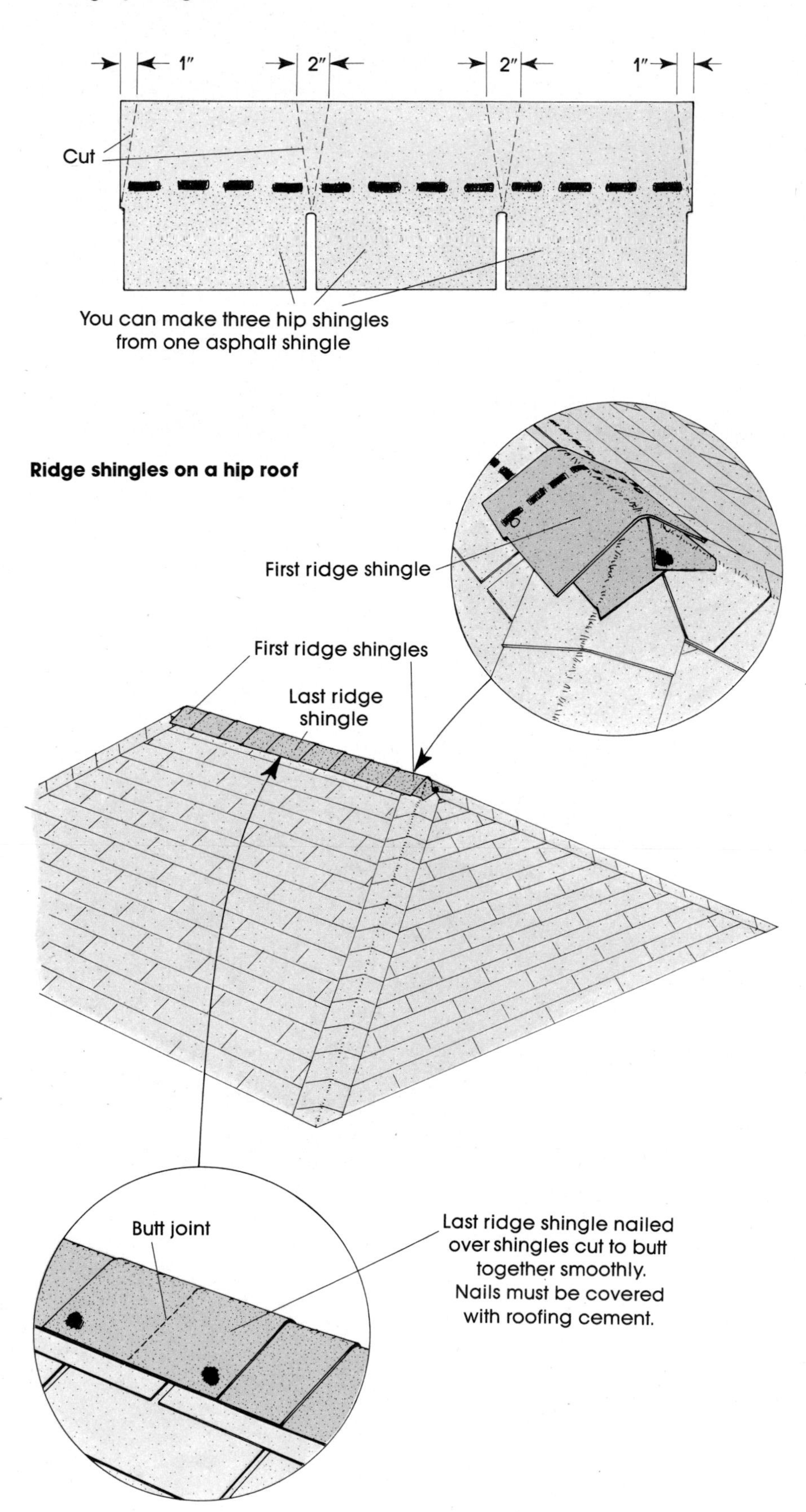

Hip Shingles

Since ridge shingles cover the top of the last hip shingle, hips should go on first. You can either make your own hip shingles by cutting them from full shingles, which is tedious, or you can buy them ready-made. If you cut your own, cut as illustrated for a really smooth fit. Put the first shingle in place at the eave, then trim along the roof edge, allowing a ½-inch overhang. Then tack a hip shingle temporarily in place at the top. Snap a chalkline between the top and bottom along one edge to keep the hip shingles straight. Remove the top shingle. Apply the hip shingles from bottom to top, putting the nails 1 inch in from the edges just below the adhesive. Use the hatchet gauge to give each hip shingle a 5-inch exposure.

Do the same on the other hips until you reach the ridge. Trim the last hip shingles so they will meet smoothly in the center rather than overlap (see illustration). Now cut a ridge shingle (they are the same as hip shingles) about 4 inches up the center, fit one split end over each hip, and nail down. Cover each nail head with a dab of roofing cement.

Ridge Shingles

Generally, ridge shingles should be applied with the exposed ends facing away from prevailing winds. However, for a hip roof, start from the hips and work toward the center. For either type of roof, put a shingle in place at each end and snap a chalkline along one edge to keep the shingles straight. Use a 5-inch exposure and put nails 1 inch from the edges and just below the adhesive.

When shingles meet in the center, trim the final shingle to fit, then cap the joint with a shingle that has had the top portion trimmed off. Nail at each corner and cover the nail heads with roofing cement.

When shingling a ridge from one end to the other, the last shingle should be trimmed and then capped in the same way.

ASPHALT SHINGLES

Shingling a Dormer Roof

Dormer roofs should be shingled when the courses on the main roof reach the eaves of the dormer roof. They are shingled in a standard manner, and the valleys protected by either a full-lace or a half-lace valley, as previously described.

Dormer ridge shingles. Dormer ridge shingles should be applied before you carry the main roof shingles across above the dormer ridge. When applying ridge shingles to a dormer, start from the outer edge and work toward the main roof. When you reach the main roof, split the top of the shingle and carry it at least 4 inches up the main roof. Shingles coming across on the main roof should lap that junction.

Tie-Ins

Tying-in a roof is the process of working around a dormer so that the cutouts between the tabs are vertically aligned, without a break, on both sides of, and above, the interruption caused by the dormer.

To do this, shingle up the roof toward the ridge while at the same time extending the courses toward the dormer. Meanwhile, carry the lower courses beyond the far side of the vertical dormer wall. As you reach the dormer roof, shingle it first, then bring the main roof shingles over to it and complete the left-side valley, as described on pages 34–35.

Now for the actual tie-in: Carry the course immediately in line with the top of the dormer roof to a point about four shingles beyond the right side (left side, if you're left-handed) of the dormer roof. Nail only the tops of the shingles so the course that is eventually brought up to it can be slipped underneath.

Continue roofing above this line all the way to the ridge. Now, using the cutouts in these upper courses as guidelines, snap a chalkline from the ridge to the eave near the right edge of the dormer. Move over 36 inches and snap another one. Use these as guidelines as you bring the shingle courses up the right side of the dormer. Slip the tops of the last course

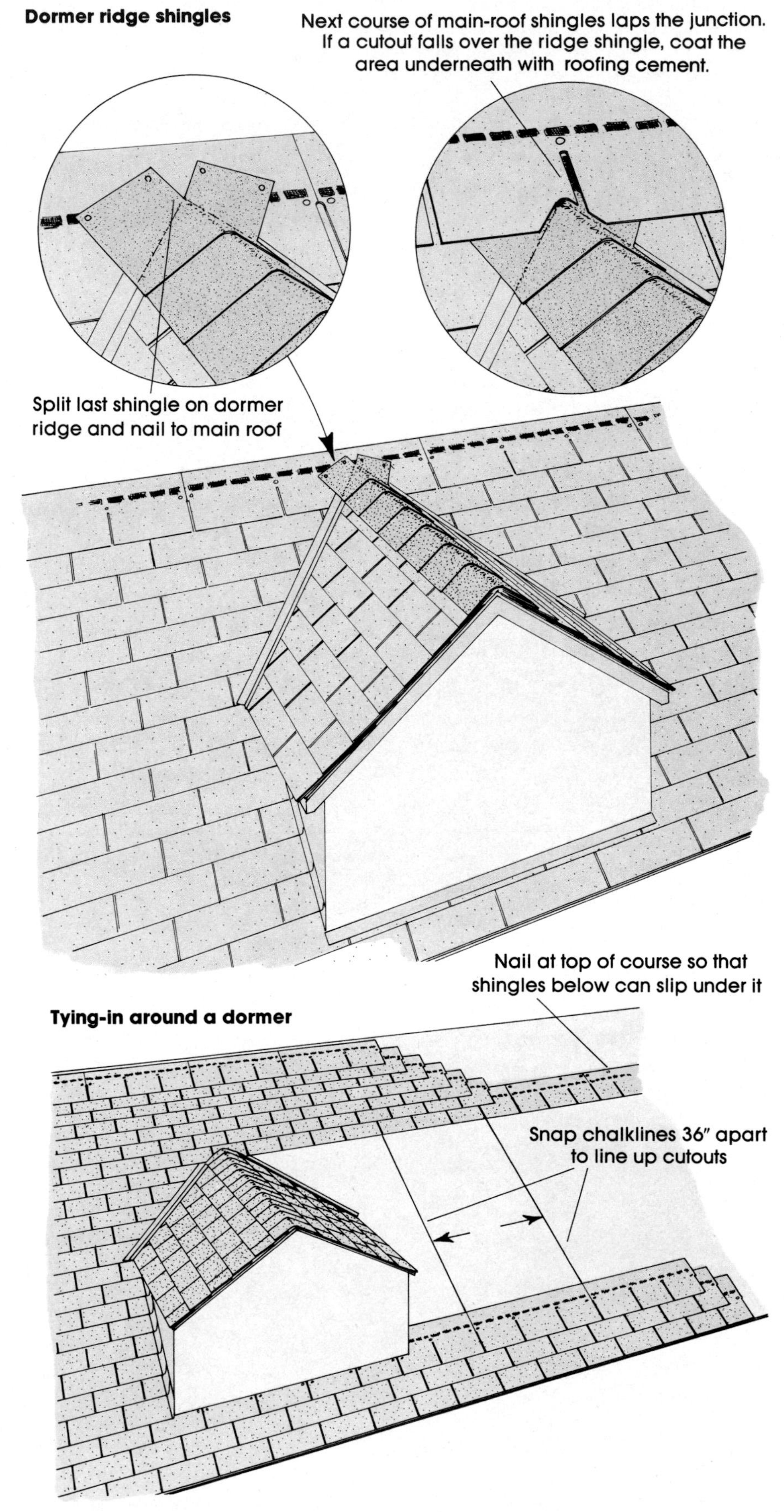

under the tabs of the course that is in line with the top of the dormer roof. Generally, as you reach the right side of the dormer, you will have to cut a shingle an irregular amount so it fits against the dormer wall and the cutout is in line with the chalkline. The second chalkline 36 inches away is a means of double-checking your work.

It's a good idea to snap horizontal lines across the roof on the right side of the dormer to keep that side aligned.

When measuring from the eaves for these horizontal lines, be sure to add the amount of shingle overhang at the eaves.

New Roofing Over Old

Covering asphalt shingles. Roofing over an existing asphalt roof adds insulation to your roof, eliminates the need for roofing felt, and saves you the time and trouble of tearing the old one off. But if you feel the old one is too irregular for a smooth new finish, tear it off. You are going to live with the new one 20 years or more and you will want it done right.

Any irregularities in the existing roof must be repaired in order to produce a smooth roof. Any warped or bent asphalt shingles should be split and nailed flat. Missing shingles must be replaced so there won't be a sag in that spot.

When covering an existing roof of asphalt shingles, it is easiest to match the shingling pattern already on the roof.

The first step in roofing over an existing asphalt roof is to apply the starter strip along the eaves. Measure the shingle exposure on the existing roof (commonly 5 inches or 6 inches). Using the new shingles, make the starter strip that width. Cut the shingles to match the exposure in the old roof, and nail the first one with the adhesive strip adjacent to the eaves, as illustrated. If the existing shingle does not extend far enough out to spill water into the gutter, cut the starter shingle wide enough to do so. Cut enough starter shingles to go along all the eaves. Discard the tab ends.

On a 5-inch-exposure roof, remove 2 inches from the top of a shingle and butt it up against the bottom of the third course of existing shingles. This 10-inch-wide shingle covers the starter course and the second course on the old roof. Cut enough of these 10-inch-wide shingles to run the length of the eaves.

For a 6-inch exposure, apply a full shingle to cover the 6-inch-wide starter strip and the second course.

For wind resistance, put a dab of cement under each tab of this first course.

Now cut 5 inches or 6 inches—depending on the shingling pattern on the old roof—off the rake side of the next shingle and butt it up against the bottom of the fourth course of existing shingles. You will note that this second course leaves only a 3-inch exposure on the first course. This is not a problem because, in most cases, it cannot be seen from the ground.

For all remaining courses, cut and apply just as you would for a new roof. The exposure follows automatically, since all new shingles are butted up against existing courses. Remember to use 1¾-inch nails.

Covering tar and gravel. As explained on page 28, it is preferable to remove the built-up roof. However, if you prefer to roof over it, follow the instructions above.

Covering wood shingles. The directions for covering asphalt shingles also apply to covering wood shingles.

Installing asphalt shingles over an old asphalt roof

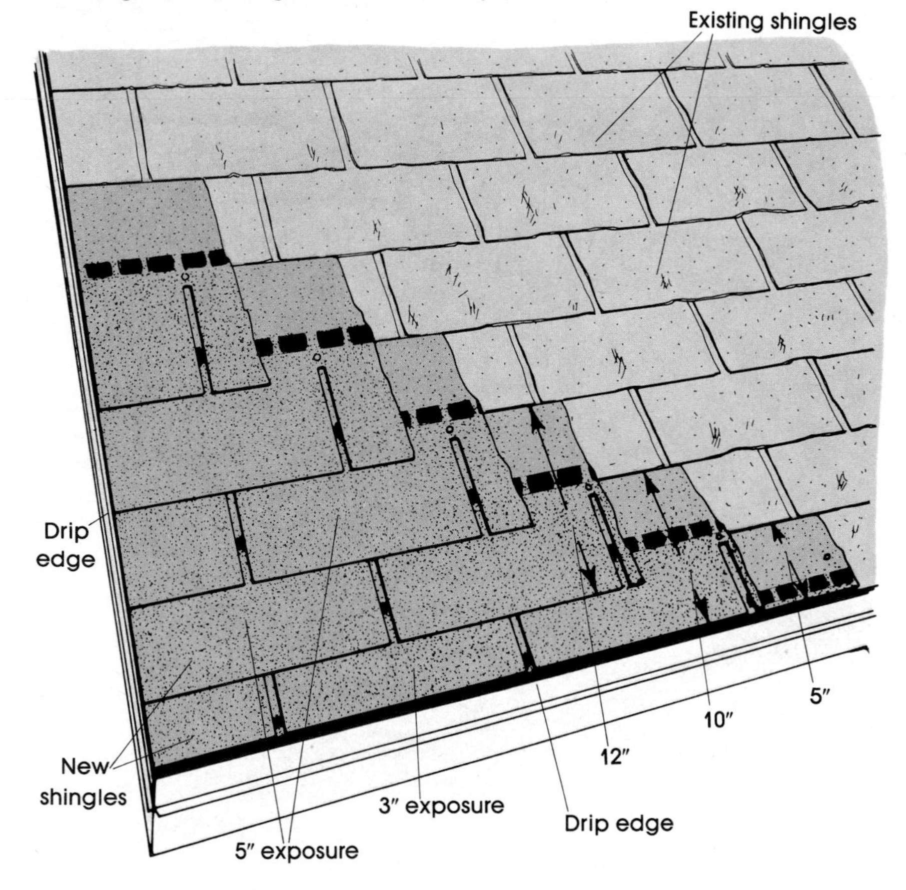

Shingling an Extra-Steep Roof

A roof is considered extra-steep if it is 60 degrees or more, or 21 in 12. Such roofs—which may be found, for instance, on mansard-style houses—require some special attention. The problem is that the self-adhesive strip does not function properly on such steep roofs.

To correct this, apply a quarter-size spot of roofing cement under each tab and press in place. Do not apply an excess amount as it may cause the shingle to blister.

SHAKE ROOF

As described on page 22, shakes come in 18-inch and 24-inch lengths, and in medium and heavy grades. There are three basic types: taper-split, hand-split and resawn, and straight-split. The first two types are thick at the butt and taper to a thin top. Straight-split shakes are equal in thickness throughout their length and are not suitable for residences because they are so bulky.

Common exposures for shakes are 7½ inches for 18-inch shakes and 10 inches for 24-inch shakes. These exposures provide the standard two-ply coverage. However, you will have a markedly better roof with three-ply coverage, which means giving 24-inch shakes a 7½-inch exposure, and 18-inch shakes a 5½-inch exposure.

As a general rule, shakes function best on roofs with at least a 6 in 12 pitch, particularly in wet and humid climates.

Sheathing

Shakes are normally laid over spaced 1 by 4s or 1 by 6s, which are referred to as spaced sheathing (see page 31). However, because the shakes' irregularities allow air to circulate under them—essential for wood roofs—shakes can also be laid over solid sheathing, such as plywood, or over existing composition roofs. Solid sheathing should be used in areas that get wind-driven snow.

The 1 by 4s for spaced sheathing should be spaced on center the same distance as the shake exposure. If you are going to have a 7½-inch exposure, nail the first board along the ends of the rafters to support the first-course butt ends. The second and all succeeding sheathing boards are then centered 7½ inches up from the rafter ends.

To speed up this process, nail on the first two courses of spaced sheathing, measure the gap between them, cut some blocks that width, and use them as spacers instead of measuring for each sheathing board.

Sheathe the top 18 inches of the roof solidly. This will allow you to adjust the exposure to make the last course come out even at the ridge.

Roofing Felt

Before applying the roofing felt, nail a drip edge along the eaves. After the felt is laid, nail drip edges along the rakes.

Use 30-pound roofing felt. The first strip laid along the eaves is 36 inches wide; all others are 18 inches wide. If you can't buy 18-inch-wide felt in your area, cut a full roll in half by cutting round and round it with a circular saw. Use an old blade or a Carborundum blade. Nail the top edges of the felt strips into the sheathing every 6 feet or so. You can either lay all the felt you will be able to cover in one day, which is preferable, or put down a new strip above each shake course.

The first strip of 18-inch felt is laid twice the exposure distance from the butts of the starter course. Thus, if you are using 24-inch shakes with a 7½-inch exposure, and the butts overhang the eaves 2 inches, then the bottom edge of the first 18-inch strip would be located 13 inches from the edge of the eaves. Right? The bottom edge of each succeeding strip should be laid out 7½ inches higher than the bottom edge of the previous one. Lay the top pieces across the ridge. At the hips, weave the strips together in overlapped joints.

Three kinds of shakes

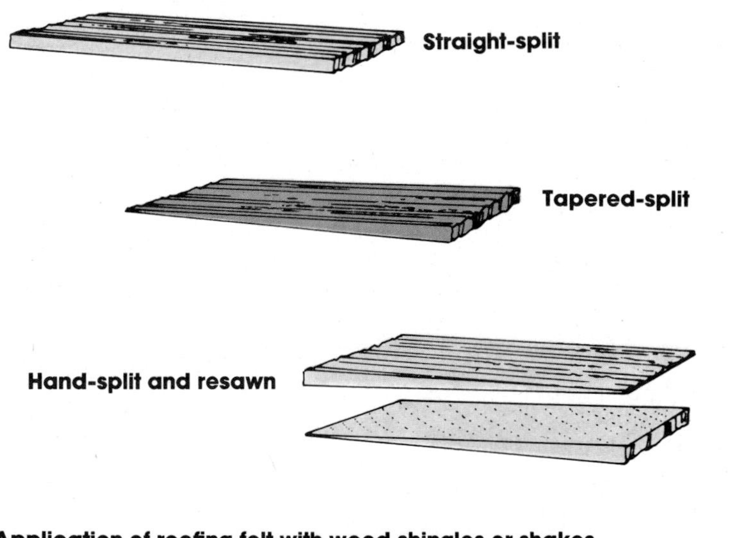

Application of roofing felt with wood shingles or shakes

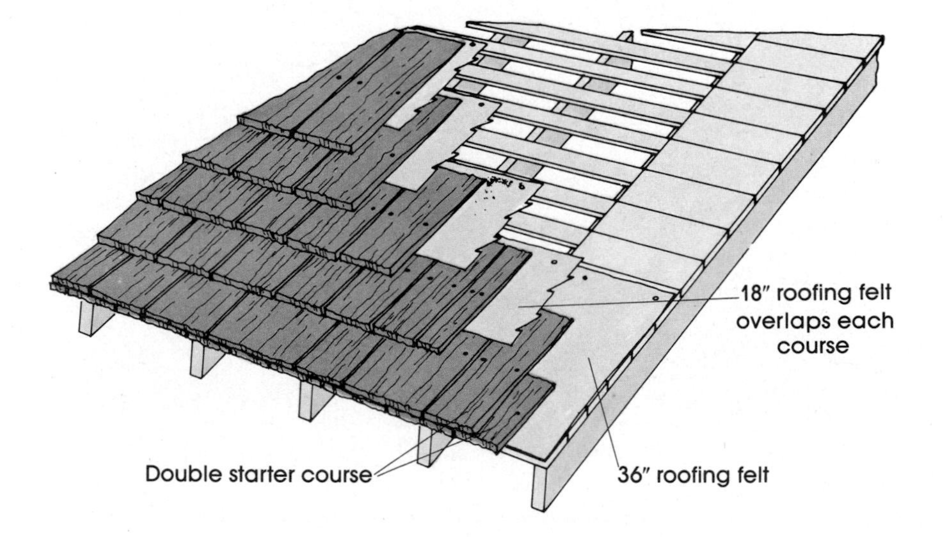

Applying the Shakes

The starter course is laid first, with the butt ends overhanging the eaves by about 2 inches. Extend the edges of the shakes over the rakes ¼ inch to ⅜ inch. Place shakes with straight, smooth edges along the rakes for a neat and professional appearance.

Slip top ends of the starter and first course of shakes under the first strip of 18-inch felt, and the top of each succeeding course under the felt above. On a properly applied roof, you should see only felt from underneath and no shakes.

Space the shakes ½ inch apart to allow for expansion. Use 7d galvanized nails placed 1 inch from each side and 2 inches under the succeeding course. Nails should be long enough to penetrate the sheathing at least ½ inch. If a shake splits, consider it two shakes and put a nail on each side of the split.

Use your hatchet as an exposure guide by marking the handle 7½ inches or 5½ inches down from the top. The joint between two shakes in one course should never be closer than 1½ inches to a joint below or above it. In addition, no joint between two shakes should be directly above another in the two courses below.

As you approach the ridge, lay out three or four courses temporarily and adjust them slightly so the final course exposure will be comparable to the others. When you nail the final two courses at the ridge, the ends will extend above the ridge. Snap a chalkline flush with the ridge and cut all of them at once with a circular saw.

Check to be sure that your work is straight by measuring, every three or four courses, the distance from the eaves to the shakes at both ends of the roof. If necessary, snap a chalkline to straighten out the next course.

Valleys. Use 20-inch-wide W-metal flashing in the valleys. To keep shakes along the valleys straight, use a 1 by 4 as a guide. Place the board in the valley, one side flush against the dividing ridge in the middle. This will provide an ample 3½-inch space for runoff on both sides of the valley.

Shakes on the left side of the valley have to be cut individually as you arrive at the end of each course. Lay the last shake at the valley over the 1 by 4, score a line across the shake, and cut.

Installing a valley on a shake roof

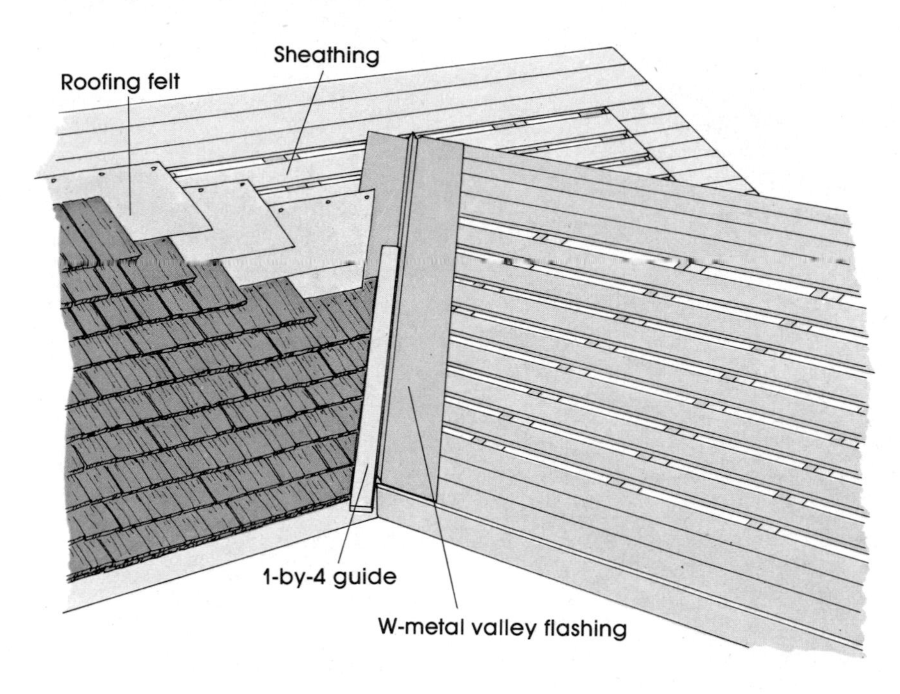

Wood-Roofer's Helper

Wood roofs are much more slippery than asphalt roofs. If you are working on a roof with a pitch of 8 in 12 or more, you can use a homemade roofer's seat to keep yourself from sliding off.

Make this box from scrap plywood or 1 by 12 lumber. The base of the box should be cut at the same angle as the roof. An easy way to do this is to place the board against the roof rake, level the top, and draw a line inside the board along the edge of the rake. The base should be about 20 inches long and the back a full 12 inches high. Make the seat about 12 inches wide. Cover the bottom with four strips of plywood or some 1 by 4s. First drive six roofing nails through each board, then nail them to the box with nails pointing up. When you turn the box over on the roof and sit on it, the nails will bite in and prevent the box from slipping.

Roofer's seat

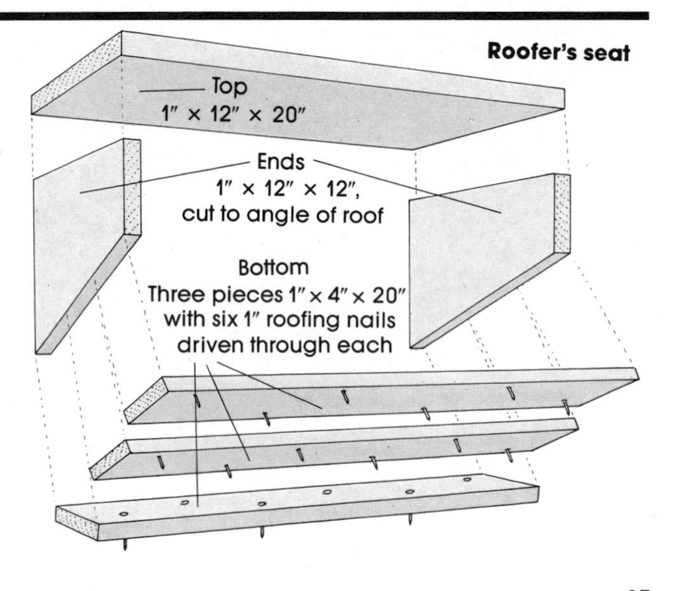

SHAKE ROOF

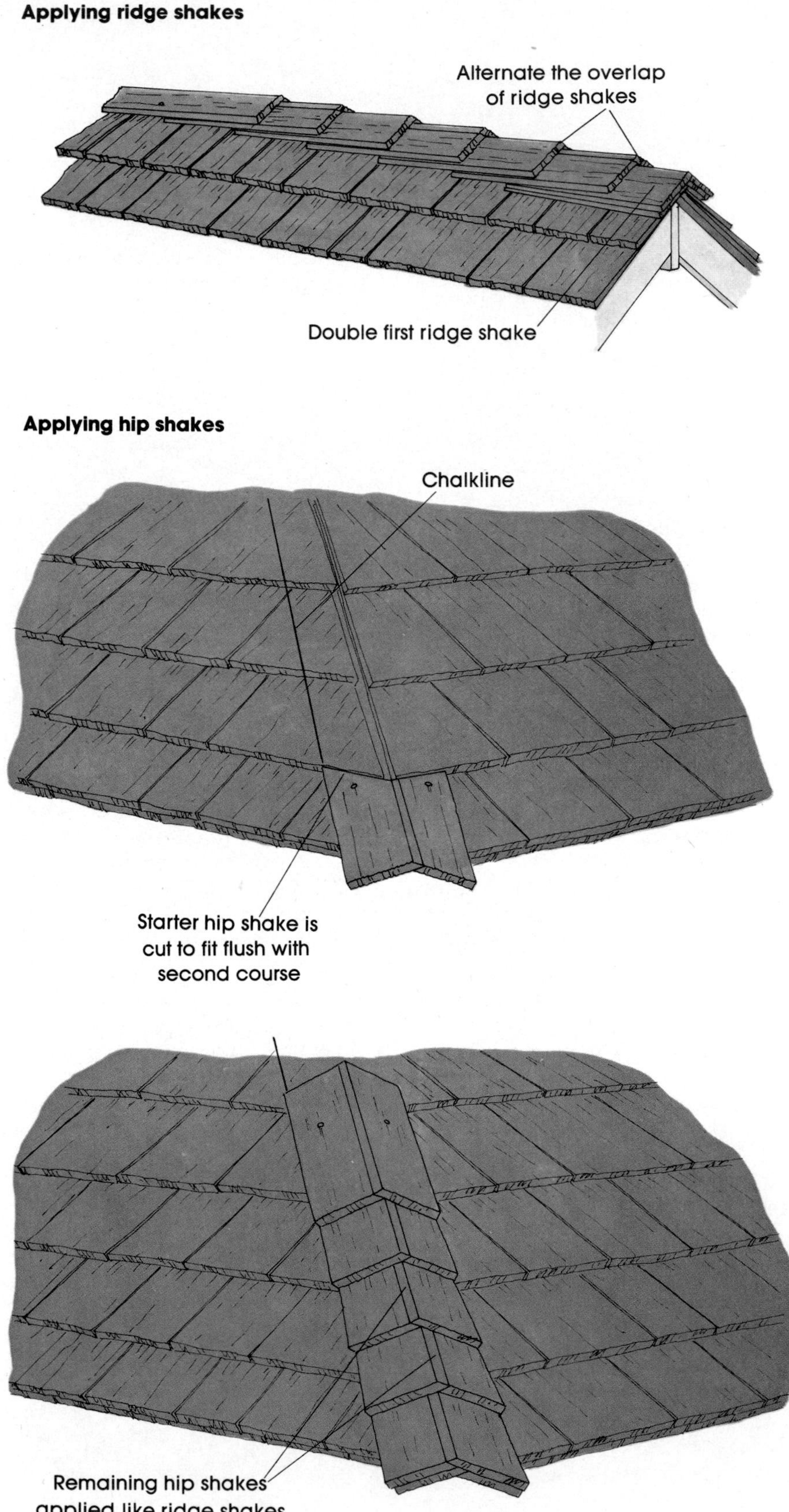

Because the right side of the valley is a starting point for courses, it goes much faster. Position the 1 by 4 and lay out the bottom shake. Mark and cut, then use it as a pattern to cut all the other shakes, several at a time. Select broad shakes for this.

Hips and ridges. Hip and ridge shakes are factory-prepared with mitered edges. Note from the illustration how they are applied, alternating the direction of the miter joints. Hip and ridge shakes are applied with nails long enough to penetrate the deck at least ½ inch.

As with shakes next to valleys, shakes running up the left side of a hip must be cut individually as you reach the hip. Use a straight board running up the hip to mark each shake. On the right side, cut the bottom shake at the proper angle and use it as a pattern to cut the others.

When applying the miter-edged hip shakes, snap a chalkline between the bottom hip shake and one placed temporarily at the top to guide your work. Apply a double hip shake at the eave. For a smooth fit, cut the starter hip shake even with the butts of the second course, as illustrated. Then apply the first hip shake over that and continue up the hip. Use 10d galvanized nails on hips.

At the top, trim the inner edges of the hip shakes where they meet each other and trim the tops flush with the ridge.

Vents. When applying shakes around vents, carry the course to the vent pipe, then use a keyhole saw or saber saw to notch the edges for a close fit on each side. Slip the flashing over the pipe, then cut two layers of 30-pound felt to fit over the flashing and extend out a foot on each side, as shown. The course above the vent can be notched at the butt if it is too close; if it is too far away, one shake can be dropped down a few inches out of line to frame the vent. It should be about 1 inch from the vent pipe. Use a broad shake here.

Next to a vertical wall, as along a dormer, use step flashing as described on page 57.

INSTALLING A

WOOD SHINGLE ROOF

Shingles are generally made from western red cedar. They come in lengths of 16 inches, 18 inches, and 24 inches, and are graded 1, 2, or 3. Only number 1 grade is free of knots and sapwood (the outer soft wood of a tree).

In many respects, applying shingles is the same as applying shakes, so refer to the preceding section as well as these directions.

Like shakes, shingles must be applied close enough together to provide triple coverage for a long-lasting roof. Use the following table as an exposure guide:

Pitch	16″	18″	24″
3 in 12, 4 in 12	3¾″	4¼″	5¾″
4 in 12 or more	5″	5½″	7½″

Do not apply shingles to roofs with less than 3 in 12 pitch.

Use 3d galvanized nails on 16-inch and 18-inch shingles; use 4d nails on 24-inch shingles. Always apply two nails per shingle, within ¾ inch of each side and 2 inches above the butt line of the succeeding course. It is important to have the sheathing spaced properly so you won't have to nail higher to hit the boards.

Sheathing

Shingles, because they are resawn and thus lie flat, need good ventilation. They are most commonly laid over spaced sheathing, although solid sheathing is sometimes used for structural purposes.

Sheathing should be spaced center-to-center, the same as the shingle exposure. Place the first sheathing board with the bottom edge flush with the rafter ends. From there, measure up the exposure distance and center a sheathing board there. Measure the gap between boards, cut some spacers, and complete the sheathing. Sheathe the final 18 inches solidly so the last few courses can be adjusted slightly to come out even.

Drip edges are not used with wood shingles.

Applying the Shingles

Lay the starter course along the eaves with butts extending 1 inch over the edge. Let them hang over the rake ¼ inch to ⅜ inch. Apply the first course over the starters with no joints closer than 1½ inches to one below. Leave ⅙ inch to ¼ inch spaces between shingles to prevent their buckling when wet.

Spend time to ensure that the second course is perfectly straight by snapping a chalkline. Use your hatchet as an exposure guide for succeeding courses and check your alignment every three or four courses.

As you approach the ridge, lay out some shingles in the final courses to see how they will fit. Adjust each one by 1 inch or less to make them come out nearly even. Let the ends extend over the ridge, then snap a chalkline and cut flush with the ridge top.

Applying shakes around vents

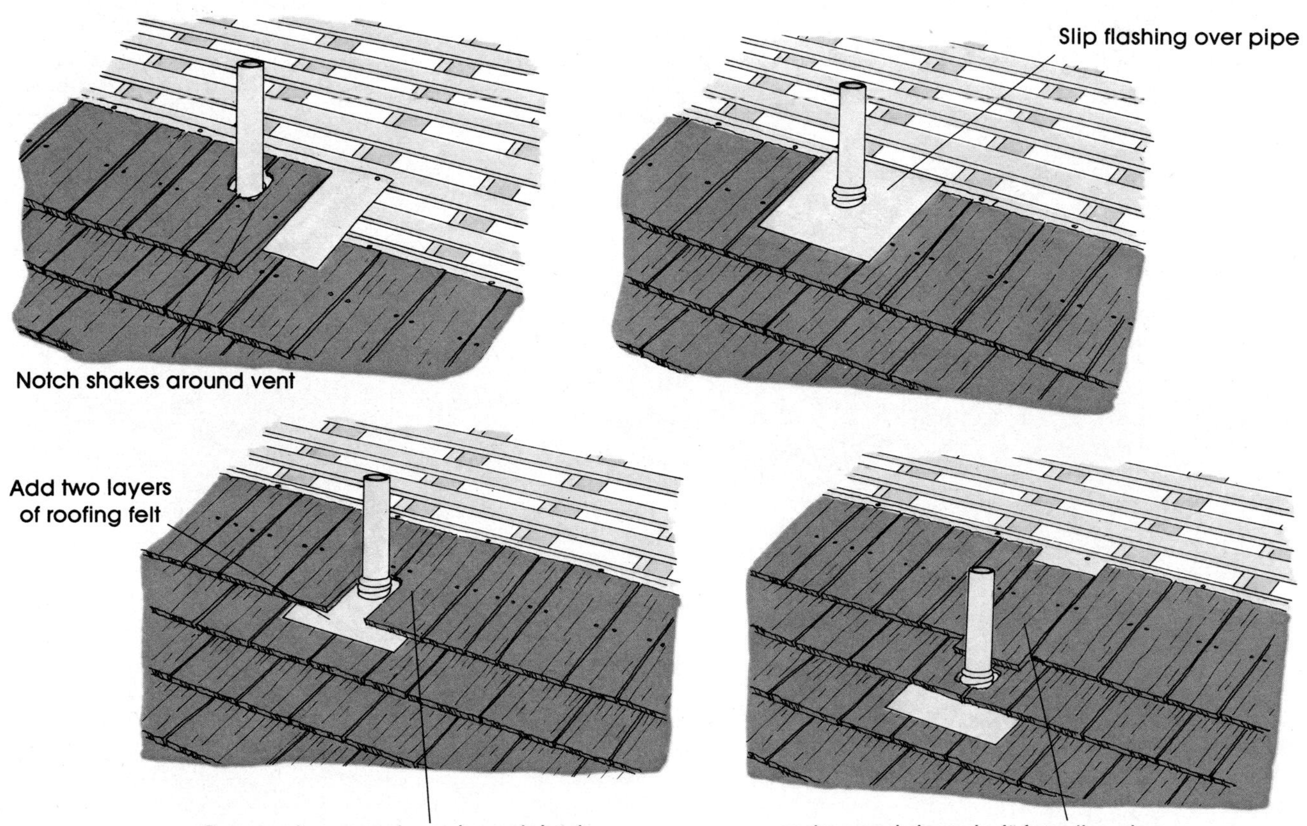

WOOD SHINGLE ROOF

CONTINUED

Valleys, hips, and ridges. The shingling process for valleys, hips, and ridges is just as described for shakes on pages 45–46. The one difference is that lengths of kraft paper (used for flashing around windows) should be laid over the hips and ridges before the shingles are applied, as an extra waterproofing precaution.

Preparing for hip and ridge wood shingles

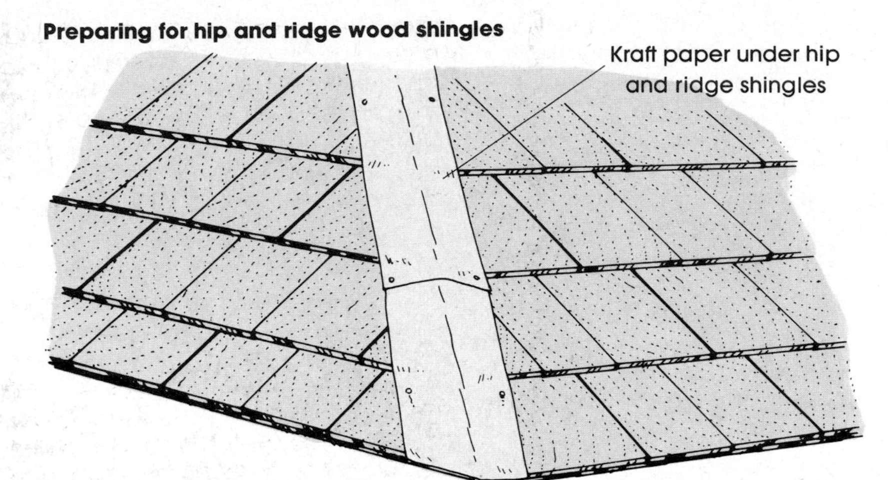

Preparing to cover old wood shingles with new ones

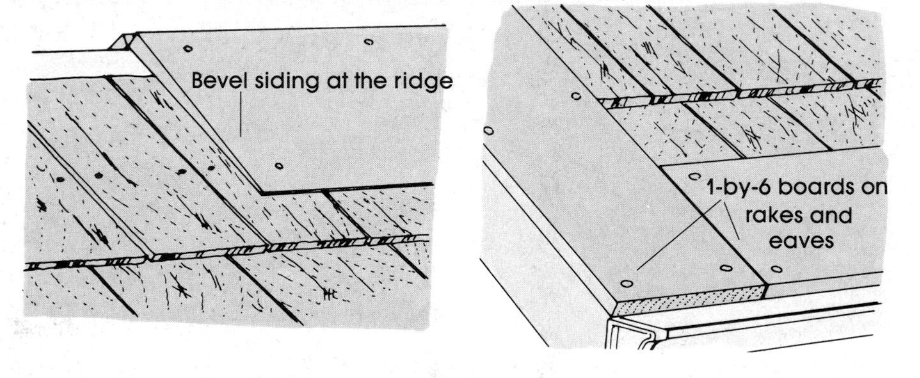

Preparing to cover an old asphalt roof with wood shingles

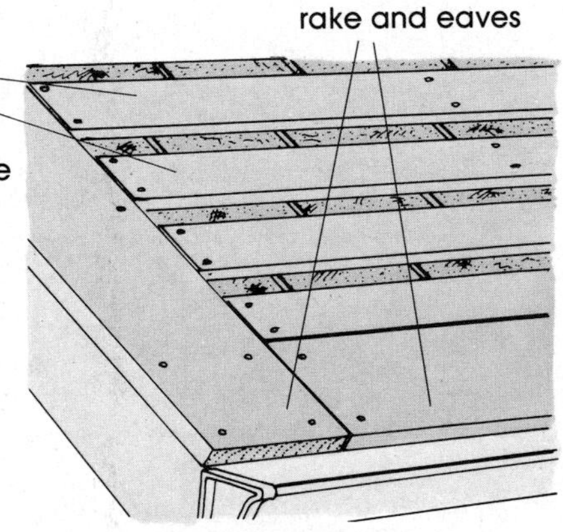

New Shingling Over Old

Covering wood shingles. If the old wood shingles on a roof are in reasonably good shape, they can be roofed over with new wood shingles. Even so, some preliminary steps are necessary. First, nail down any curled or warped shingles. If one won't go down, split it, pull out the pieces, and slip a new shingle in to keep the surface even.

Shingles along the eaves, rakes, and ridges should be removed and replaced with a 1-by-6 board. Measure 5½ inches back from the edges of the rakes and eaves, then snap a chalkline as a cutting guide. Set the blade on a circular saw just slightly beyond the shingle depth and cut. Use an old saw blade or carbide-tipped blade, since you will hit some nails. Follow the same procedure on both sides of the ridge.

Sweep the roof, then nail down the 1 by 6s along the eaves and rakes. At the ridge, use a length of bevel siding, with the thin edge on the down side. Apply new flashing in the valleys and along the eaves and rakes.

Once this process is complete, use standard shingling procedures to apply the new roof.

Covering asphalt shingle, roll-roofing, and tar and gravel roofs. Applying shingles over these roofs is quite similar to new shingling. They must be placed on spaced sheathing nailed directly to the existing roof.

For an asphalt roof, remove the shingles along the ridge and hips. No removal is necessary with roll-roofing or tar and gravel roofs. Staple lengths of window flashing paper over the gap as extra protection against leaks.

Use tin snips to trim off the edges of the asphalt shingles where they overhang the rakes and eaves.

Next, nail a 1 by 6 along the rakes, from ridge to eaves. This provides a finished edge when the shingles are in place. Nail 1 by 6s along the eaves and on both sides of the ridge, with the edges of the boards meeting at the ridge. Finally, nail a pair of 1 by 4s down each side of the valleys to provide support for the new valley flashing that must be installed. Now place spaced sheathing over the roof, with the 1 by 4s spaced the same as your shingle exposure. See page 31 for detailed instructions.

Before you start shingling, lay the valley flashing in place, as described on pages 34–35. Now shingle the roof in a standard fashion, as described above.

INSTALLING A TILE ROOF

For a roof that is ruggedly beautiful, fireproof, and long-lasting (50 years or more), consider the tile roof. The Spanish were making the distinctive barrel-shaped clay tiles in California long before it became a state, and that design is still widely seen in this country. But clay tile is both expensive and heavy—roofs must be able to handle about 1000 pounds for every 100 square feet of tile.

However, recent innovations have brought down both the price and the weight of tile. It is now widely made in lightweight concrete that weighs less than 700 pounds per square and can be applied on a standard roof built for asphalt shingles. Concrete tiles come in a variety of colors and patterns that make it possible to compliment or match the siding on your house. Roofing supply firms should be able to give you information describing the types of tiles available, including their weight per square and instructions on how to install them. One of the easiest of the new concrete tiles to apply is the flat tile, which is described here.

If in doubt about the weight on your roof, even with the lighter concrete tile, discuss it with your local building inspector before ordering.

Tile cannot be put on a roof with less than a 3 in 12 pitch without risk of rain blowing up under the tiles. On pitches of 3 in 12 and 4 in 12, it is advisable to lay a bead of caulk under the overlapping edges of the tiles to prevent rain-blown leaks.

Some flat tile is made with a built-in air pocket that provides an R-11 insulation value. This value can be markedly increased by covering the sheathing with ½-inch rigid insulation that in turn is covered with heavy-duty foil. The foil blocks solar rays during the summer and reflects heat back inside the house during winter heating months.

Preparing the Roof

The roof should be solidly decked with ½-inch exterior-grade plywood or, for an exposed ceiling, with 2-by-6 tongue-and-groove boards.

In applying roofing felt, follow the manufacturer's instructions, which may call for 15-pound, 30-pound, or none at all.

Metal drip edges should be applied along the eaves under the felt and along the rakes over the felt, in the same manner as for a standard roof (see page 38).

Valleys should be flashed with W-metal flashing at least 24 inches wide laid over 90-pound mineralized felt. Hips and ridges should be covered with a double layer of felt that overlaps at least 6 inches on both sides.

All hips and ridges must have a 2-by-2 board down the center nailed at 12-inch intervals to provide support for the curved hip and ridge tiles.

To keep water dripping off the rake tile from hitting the end rafter, a 1-by-3 spacer board is nailed along the outer edge of the rake.

The final preparation—and an important step—is to nail a 1-by-2 redwood or treated wood starter strip along the edge of the eaves. Use a stretched string alongside to ensure this eave board is straight. Fasten it with 8d galvanized nails placed every 12 inches and nailed into the rafters when possible.

Preparing for a concrete-tile roof

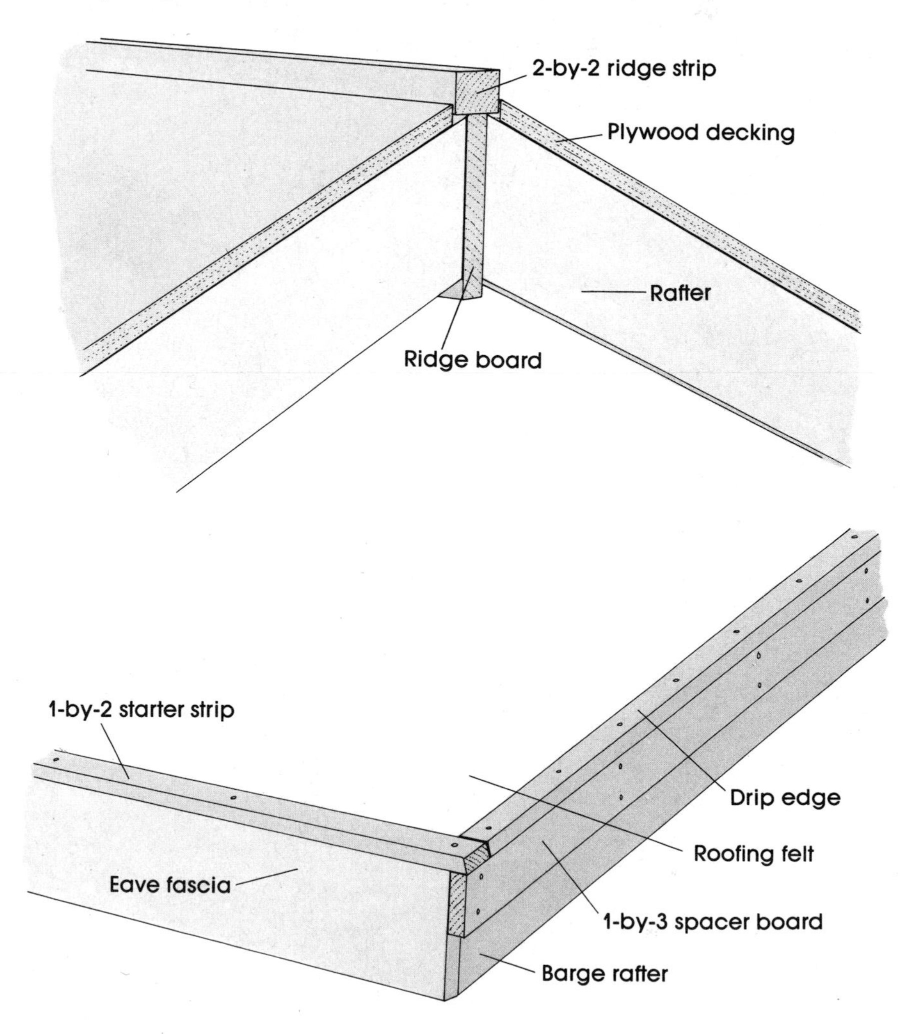

TILE ROOF

Installing the Tiles

Unlike standard shingles, concrete tile does not require a doubled first course. The 1-by-2 starter strip along the eaves serves the same purpose by raising the butt end of the first tile course.

Start each course with the tile edge flush with the outside edge of the 1-by-3 spacer board on the rake. After all courses are installed, trim tiles are laid along the rake.

The first tile course is laid along the eaves with the butt end overlapping the 1-by-2 eave strip. Each successive course simply locks into the previous one.

Two nail holes are provided near the top of each tile. Use galvanized nails long enough to penetrate ¾ inch into the roof deck.

On roofs with a pitch from 7 in 12 to 10 in 12, use the clips provided with the tiles to attach every fourth row. On roofs with higher pitches, use clips on every tile.

Valleys, hips, and ridges. Hip and valley tiles are laid out and cut in the same manner as wood or asphalt shingles, using a tile-cutting Carborundum blade in a circular saw. Use a chalkline to mark hips 1½ inches back from the edge of the 2-by-2 center strip, and mark valleys 2 inches back from center on each side. Bring the last tile in each course to the chalkline, lay a straightedge over the shingle in line with the chalkline, then mark

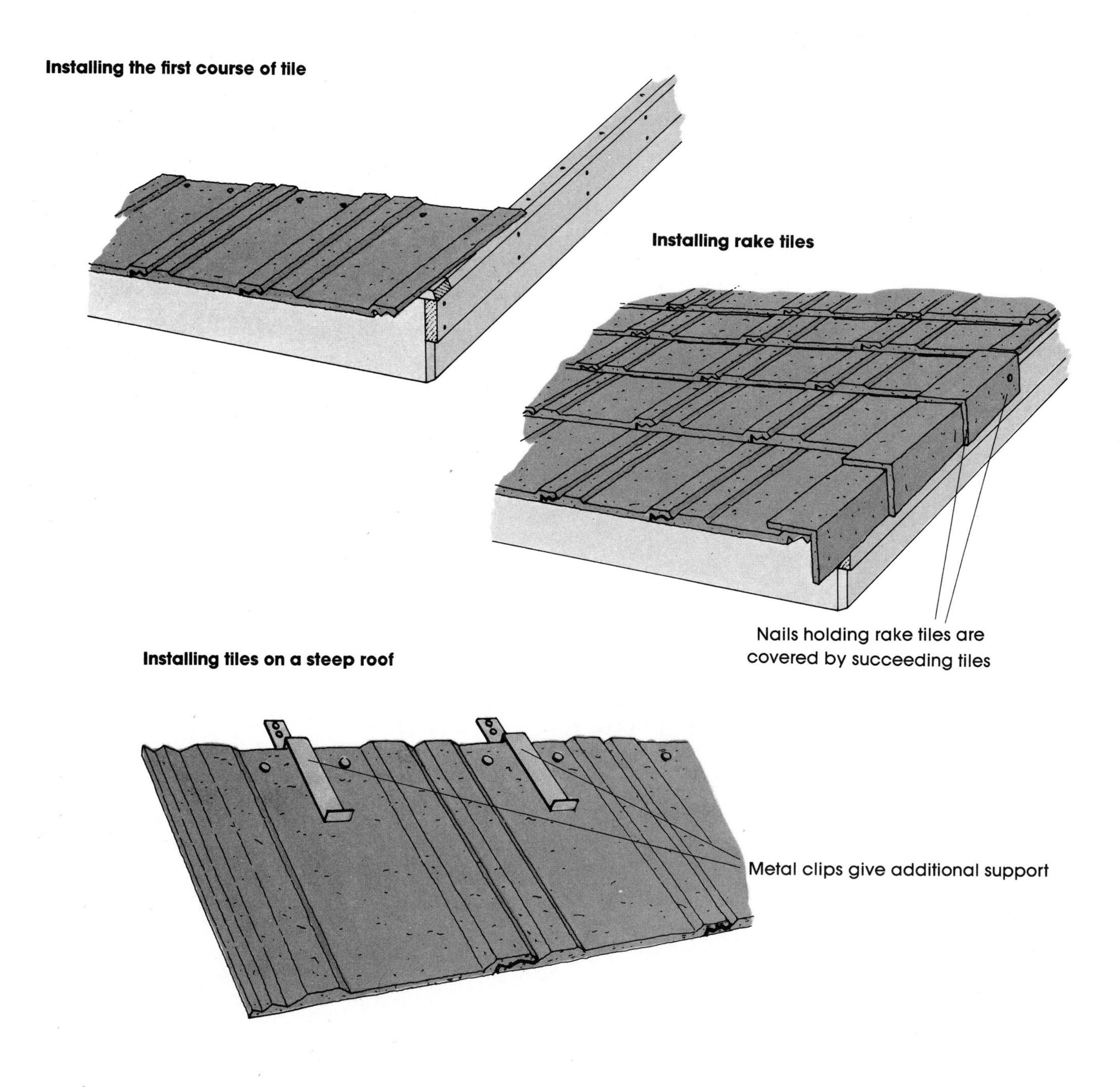

and cut the tile. Lay a thick bead of caulk under and along the edge of each tile on the valley flashing.

On ridges, carry the final courses to the top, then trim the last course 1½ inches back from the edge of the 2-by-2 center strip.

Both hip and ridge tiles must be placed on a bed of concrete that covers the tile edges on both sides of the center strip. Spread the concrete in place, then nail the hip and ridge tiles down while the concrete is still soft. Cover each nailhead with roofing cement.

Vents and chimneys. When a tile reaches a vent opening, it must be cut to fit snugly around the vent pipe. After measuring the distance to the vent and marking the tile, make the cut by raising the blade guard on a circular saw and lowering the moving blade slowly into the tile along the cutting lines. Rap the center piece with a hammer to break it out at the corners where the cut was not complete. Vent flashing for tile roofs must be specially made from lead at a sheet-metal shop. Apply the flashing in the standard manner described on page 55, then mold the soft lead to the shape of the tile. Fill any gaps between the flashing and the cut piece of tile with roofing cement.

Around chimneys, install flashing and place tiles as with asphalt shingles. For details on flashing a chimney, see pages 58–59.

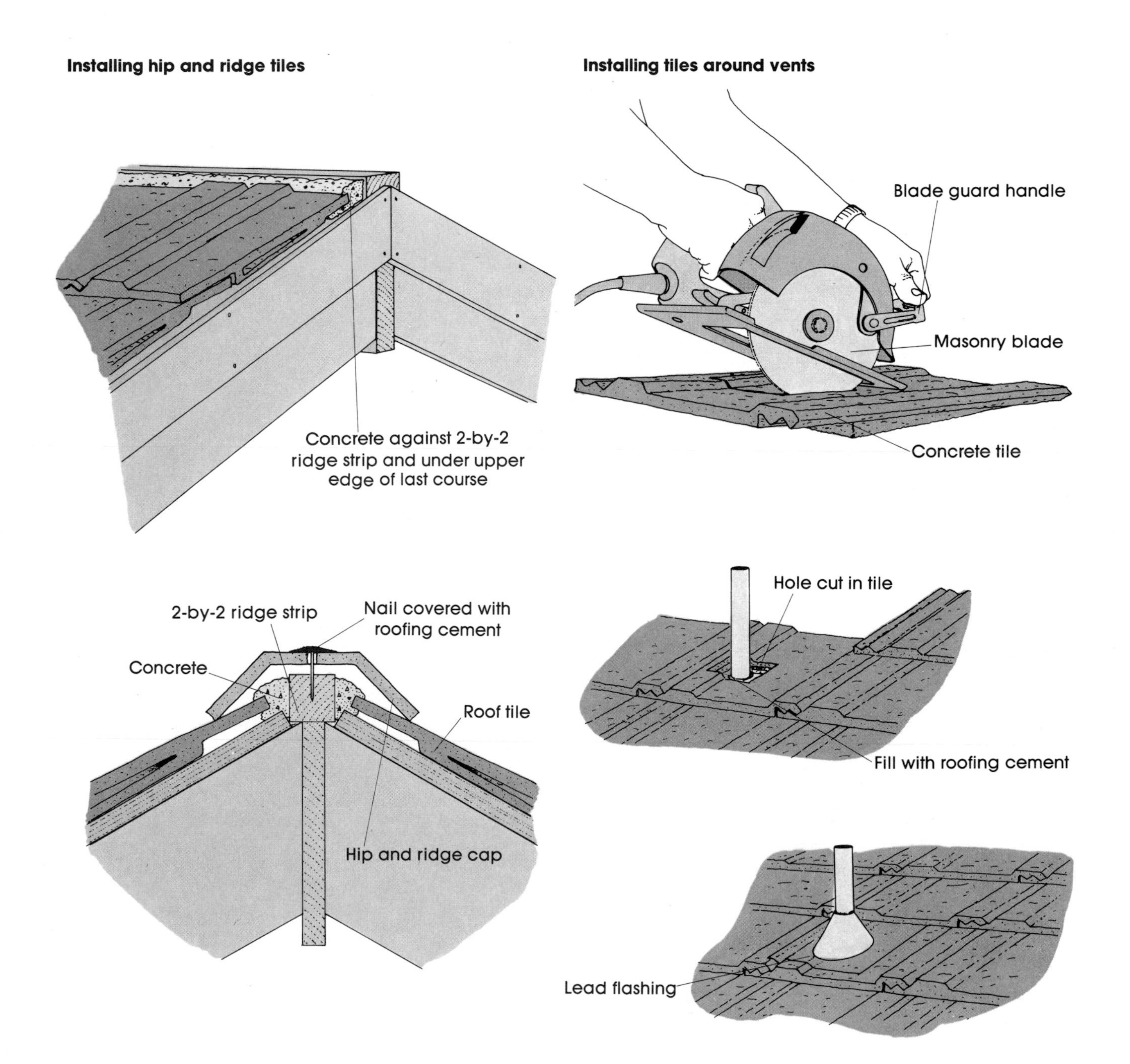

INSTALLING

ROLL-ROOFING

Roll-roofing comes in a variety of colors and weighs 70 to 90 pounds to the square. It is sold in 36-inch-wide rolls that cover one square. It can be used on roofs down to a 2 in 12 pitch with the exposed-nail method, or down to 1 in 12 with the concealed-nail method.

As a rule, because roll-roofing can crack in cold weather, it should not be applied when the temperature is below 45 degrees F. If the job must go ahead anyway, store the rolls in a warm area prior to application. If that is not possible either—life is tough—then carefully unroll them, cut lengths the width of the roof or not more than 18 feet, and let them lie on the roof until they are flat. In very warm weather, be careful that your shoes don't gouge the felt.

Apply roll-roofing over a smooth plywood deck with a ½-inch overhang at the eaves. Cut it flush with the rake edges and cover with metal drip edge.

Exposed-Nail Application

If there are valleys, these should be covered first with 18-inch-wide strips of matching roll-roofing. Seat it firmly in the valley without bending the center area so sharply that it cracks. Nail one side first with galvanized roofing nails placed ¾ inch from the edge and spaced every 6 inches. Now seat the strip in the valley and nail the other side. If you must use more than one strip, lap them by 6 inches and coat the entire area with roofing cement.

To position the top edge of the first course, snap a chalkline across the roof 35½ inches up from the eaves. Position the roll carefully, then put nails every 2 feet along the top, ¾ inch from the edge. Now nail down the rake and eave edges with nails ¾ inch from the edge, 3 inches apart. If one sheet doesn't reach across the roof, overlap the next piece by 6 inches. Nail the underneath section down first, coat with roofing cement, and nail the overlap in place.

Snap a chalkline 2 inches down from the top edge of the first strip as a guide for the next strip. Tack the upper edge of the second

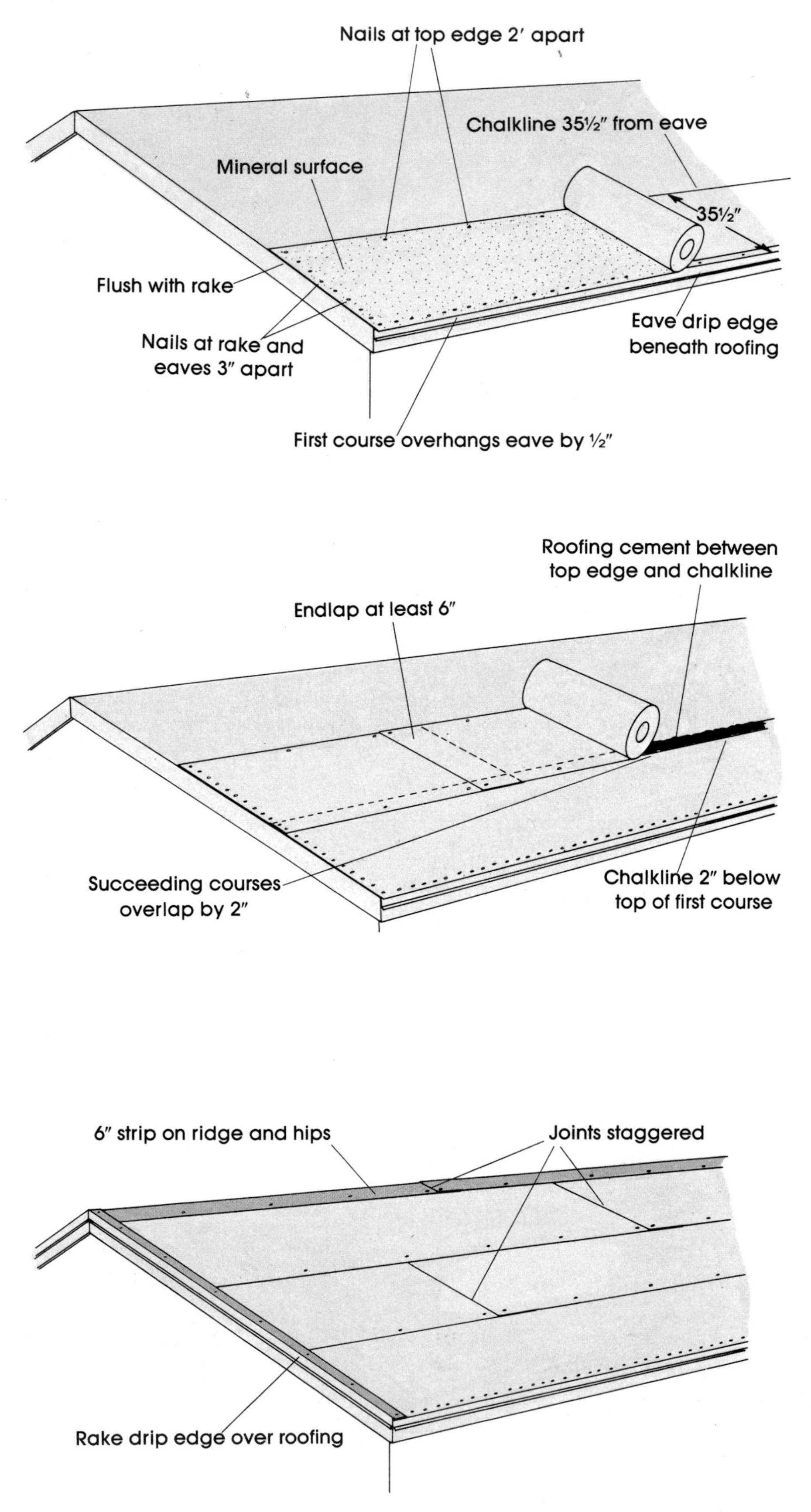

strip, then spread a 2-inch-wide layer of roofing cement along the upper edge of the first strip and nail the second course over it.

If succeeding courses must be endlapped, stagger the joints so that they are not directly above each other in adjoining courses.

Hips and ridges. Cut the roofing so it meets but does not overlap the joint. Snap a chalkline on each side 5½ inches out from the center of the hip or ridge. Spread a 2-inch-wide layer of roofing cement from each line back toward the center. Cut a strip of roofing 6 inches wide the length of the hip or ridge, gently bend it in the center to fit over the joint, and nail in place. Any endlaps must be overlapped 6 inches and coated with roofing cement.

Concealed-Nail Application

This method is used on slopes of 1 in 12 or less where water runoff is so slow that it could work under exposed nails. Since roofs with such low pitch would not have hips or ridges, in practice this method is used only on shed roofs.

First, install valley flashing as described for the exposed-nail application. Then cut 9-inch-wide strips and place them along the rakes and eave, as shown. Nail ¾ inch from the edges.

Snap a chalkline across the roof 35½ inches from the eave and position the first strip along it. Nail the top edge only, with nails ¾ inch from the edge every 4 inches.

Lift the edges along the rakes and eave and coat with roofing cement, then press the full strip into it. Any endlaps must be 6 inches wide, with the bottom layer nailed and coated with cement and the top strip pressed into it.

Position the second course on a chalkline 4 inches down from the upper edge of the first strip. Nail it every 4 inches. Coat the overlap here and along the rakes with cement and press the strip into place.

Continue up the roof in this manner. When you reach the top edge, cover it and the rakes with metal drip edging. Apply a layer of cement under the drip edges and nail in place every 6 inches.

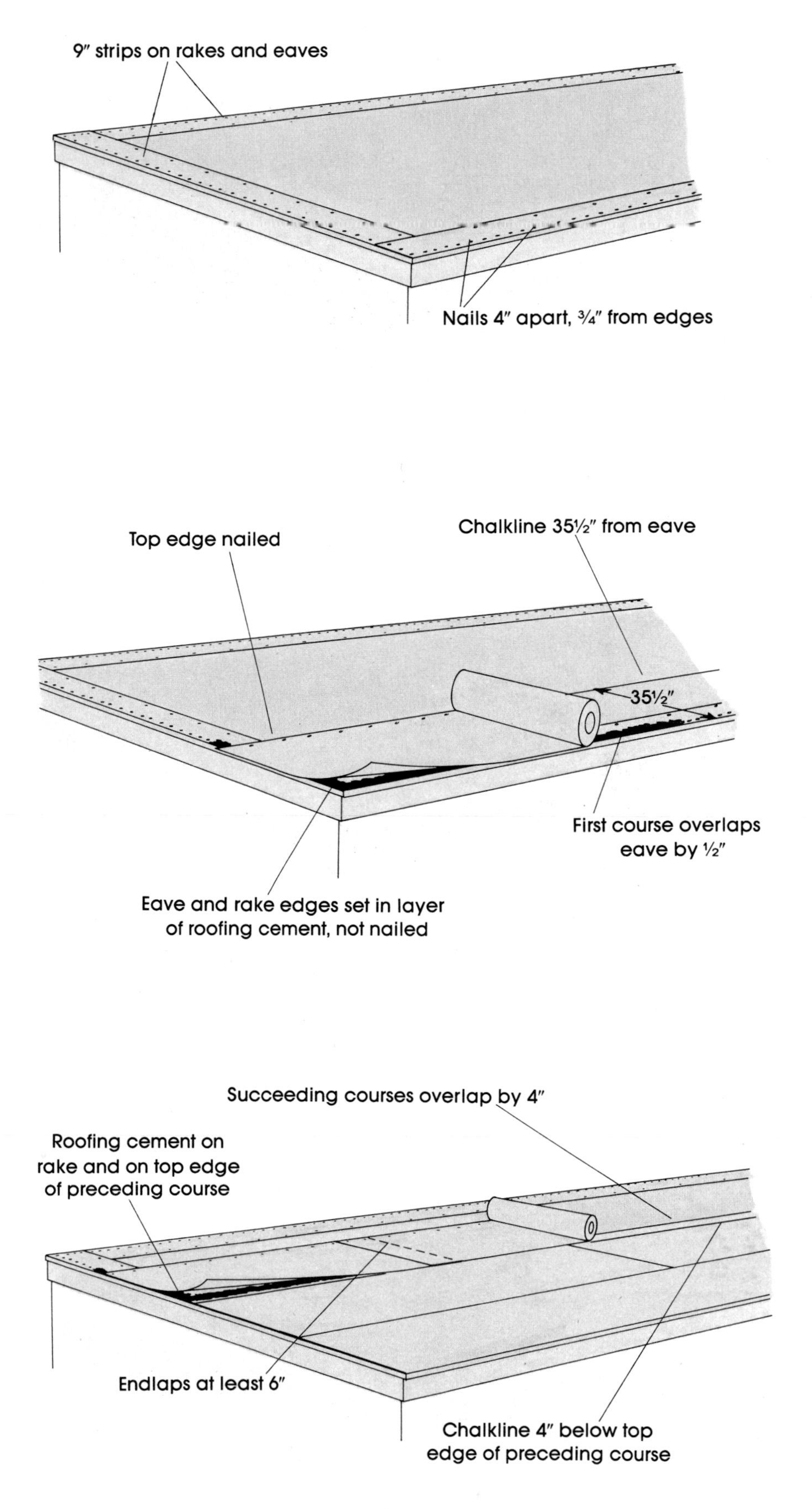

INSTALLING A

PANEL ROOF

Aluminum panels and corrugated galvanized steel roofing are excellent for farm structures and vacation cabins. Fiberglass panels, factory-treated to resist darkening by solar rays, are widely used for greenhouses and patio covers. All are applied in a similar fashion.

The lightweight and easy-to-apply aluminum panels readily shed snow, but such a roof can be noisy in hailstorms or rainstorms. Good insulation reduces this problem. The appearance of roofs can be improved by ordering factory-painted panels. Unfortunately, this nearly doubles the price, which otherwise is comparable to asphalt shingles. You can paint the roof yourself, but the metal must be cleaned first with muriatic acid. Or you can let the roof weather for one full year, then paint it with an exterior-grade metal paint.

Panels, which come in lengths ranging from 8 feet to 20 feet, are most easily installed on shed or gable roofs. Cutting for hips and valleys will slow down an otherwise fast job. Cut both metal and fiberglass with a Carborundum blade a circular saw. Be sure to wear safety goggles.

Panel roofing is nailed to 1-by-4 sheathing strips. Space these either 2 feet or 4 feet on center, depending on the roof's steepness and any possible snow load. In heavy snow country the roof should have a pitch of at least 8 in 12, with sheathing strips spaced every 2 feet.

Aluminum panels are nailed down with aluminum nails that have a neoprene washer under the head. Using steel nails on aluminum roofs will trigger a chemical reaction that destroys both metals—thus you must also use steel nails on steel roofs. Nails are always placed on top of a ridge in the panel rather than in a valley, where more water flows. Drive nails so the washer is seated firmly against the panel but causes no indentation. Panels are normally 26 inches wide, which allows a 1-inch overlap at each side. Put four nails across the panel over each sheathing board.

Steel and fiberglass panels are applied in the same fashion, but the nail holes must be predrilled. Steel is too tough to drive a nail through and fiberglass will shatter without a predrilled hole.

Installing the Panels

If your roof has valleys, install W-metal flashing (see page 34). Drip edges are not necessary.

The first panel should be placed at the bottom left corner of the roof. Allow a 2-inch overhang at the eaves and overhang the rakes by 1/4 inch to 3/8 inch. The first panel must be installed perfectly straight because all others interlock and there is little room to make an adjustment if it's crooked.

Because the panels come in such a wide variety of lengths, you should be able to install a roof with no leftover pieces. For roofs that require more than one panel from eaves to ridge, order long and short pieces and then overlap them. The overlap should be 12 inches to 18 inches, with the longer lap used on lower-pitched roofs, such as those under 4 in 12. For added protection on low-pitched roofs, place a thick bead of caulk under the bottom edge of the overlapping panel.

Ridge caps are factory-supplied to match your roof panel configuration. They commonly overlap each other by half a panel width and are simply nailed in place. There are also other factory-supplied pieces such as rake and corner trim that you may wish to apply.

A tip: If you miss the sheathing with a nail, lift the nail up slightly, put a thick bead of caulk under the washer, then press it in place.

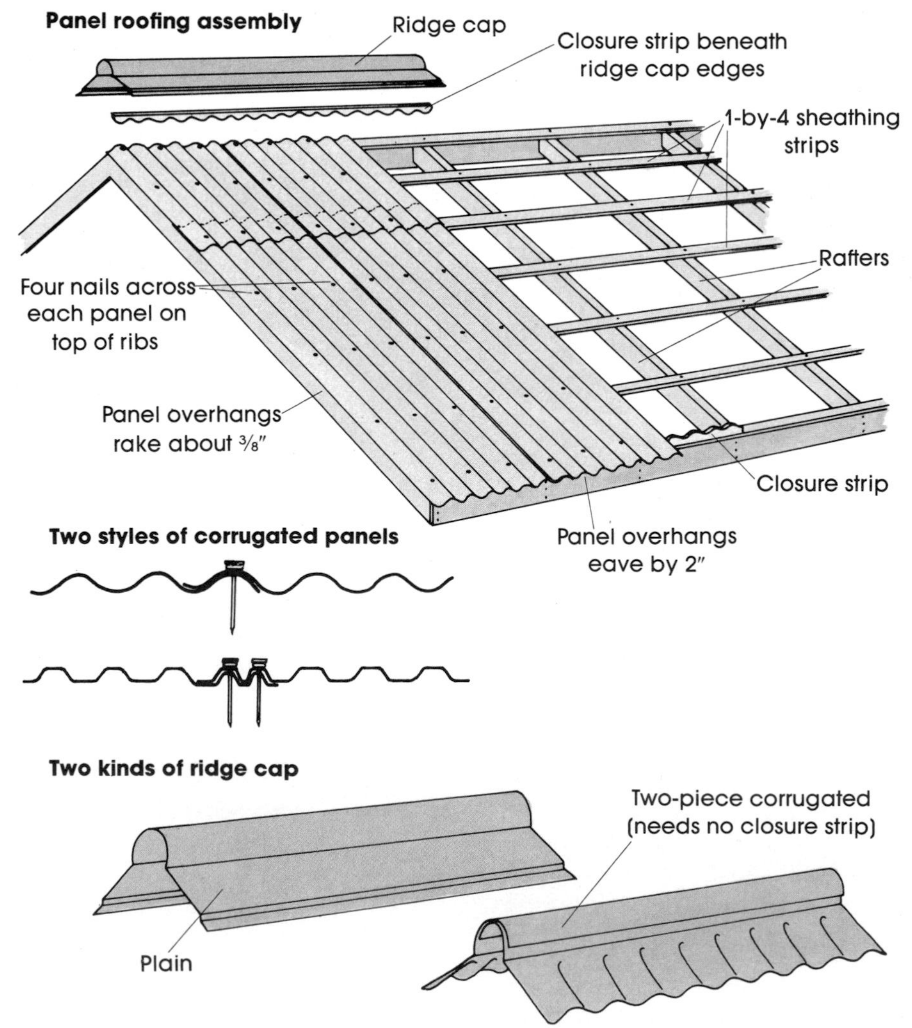

FLASHING

Flashing is used to keep water out of a house wherever something extends through the roof, such as a vent pipe, chimney, or skylight; where two roofs connect to form valleys; or where a roof meets the side of a house, such as with dormers or additions. The section that follows covers all aspects of flashing except drip edges, which are included in the instructions for installing particular roofs; valley flashing, which is done prior to the installation of the roofing material and is covered on pages 34–35; and skylight flashing, which is described in detail in the section on installing a skylight in Ortho's book, *How to Replace & Install Doors & Windows.*

The types of flashing explained here are vent, continuous, and step. All three types utilize metal flashing, which may be copper, aluminum, or galvanized tin. Here's a look at them, starting with the easiest.

Vent Flashing

Flashing around vent pipes has become quite straightforward since the arrival of the rubber sleeve. It is widely used on shake and asphalt and wood shingle roofs. Vents in tile roofs are flashed with lead; panel roofs and roll-roofing require special approaches.

Shake, asphalt shingle, and wood shingle roofs. The rubber sleeve for vent flashing is a flat piece of galvanized metal with a rubber collar in the center that slips over the vent pipe. Buy the sleeve according to the diameter of the vent, commonly 1½ inches to 3 inches.

Before putting the sleeve over the vent, bring a course of shingles or shakes up to the bottom edge of the vent pipe. If the top shingle—wood or asphalt—is hitting the pipe, notch it to fit. If that course extends well over the pipe, cut a hole in an asphalt shingle and slip it over; for shakes or wood shingles, use a keyhole saw to notch two shingles so they fit around the sides.

Slip the flashing sleeve over the pipe and coat under the metal with roofing cement.

For asphalt shingles, bring the next course across. Where a shingle meets the vent, cut it in a smooth arc about ½ inch away from the pipe. Cutting it too close will allow debris to collect there and possibly dam up water. Embed each shingle that fits over the flashing in roofing cement. Do not use an excessive amount on asphalt shingles, since it can cause them to blister in the hot sun.

For wood shingles or shakes, choose a broad shake or shingle to go above the vent. For a really fine job, use a keyhole saw to cut an arc to fit the shingle or shake around the pipe. The alternative is to back that shingle or shake away from the pipe. If the course above the vent pipe is too far away to cover the flashing thoroughly, drop the closest shingle down. It won't be noticed.

Don't forget that the lower edge of the flashing always rides on top of the course of shingles below it.

Tile roofs. Vent flashing for tile roofs must be made of lead so it can be molded to the shape of the tile. Apply like standard vent flashing but fill the gap between the flashing and the cut-out tile with roofing cement.

Roll-roofing and panel roofs. On roofs that do not have shingles or shakes under which the top part of the flashing can slip, you have to improvise. This is the case for panels and roll-roofing.

For roll-roofing, or for a vent that comes through a flat area of an aluminum panel roof, cut a hole just large enough for the vent pipe. Run the vent pipe through the hole and slip rubber sleeve flashing over it. Cover the area under the metal flashing with a thick layer of roofing cement and embed the metal in it. Nail through the metal and the roof into cross supports that you must insert between the rafters. Space roofing nails 2 inches apart all around the edge. Cover the edge of the metal flashing and the nail heads with a smooth layer of roofing cement.

For corrugated metal roofs and fiberglass, you cannot use the flashing sleeve. Pack the opening in the roof and around the pipe with roofing cement. Spread the cement above and around the pipe so it rises in a smooth sweep from the roof about 3 inches up the vent pipe.

Three kinds of vent flashing

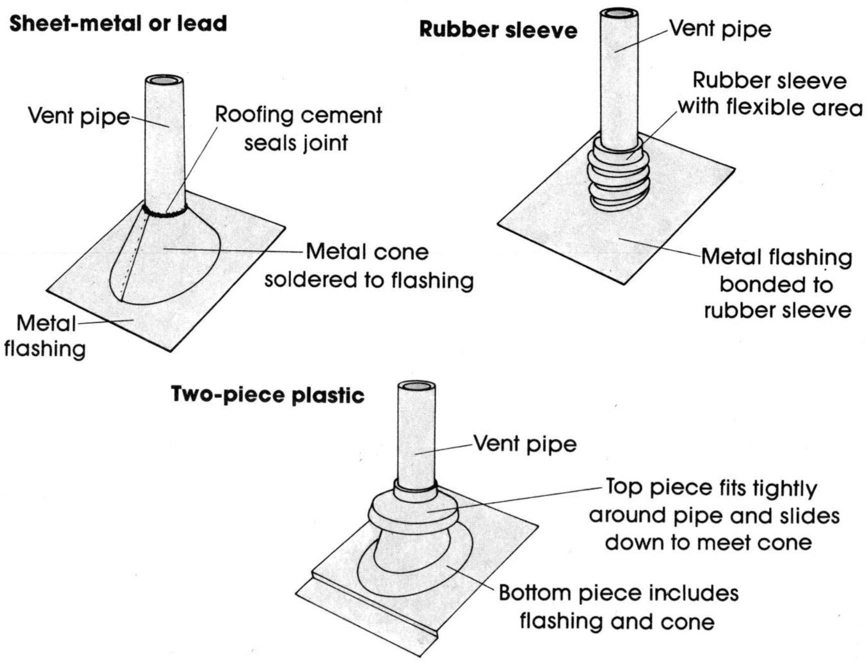

Sheet-metal or lead

Vent pipe
Roofing cement seals joint
Metal cone soldered to flashing
Metal flashing

Rubber sleeve

Vent pipe
Rubber sleeve with flexible area
Metal flashing bonded to rubber sleeve

Two-piece plastic

Vent pipe
Top piece fits tightly around pipe and slides down to meet cone
Bottom piece includes flashing and cone

FLASHING

Continuous Flashing

This type of flashing is typically used for vertical walls where a roof meets the front wall of a dormer, or where a shed roof is attached to a wall. Plan the last two or three courses before the vertical wall so the final one, which will go over the flashing, will be at least 8 inches wide. Install the flashing before applying the last course.

The flashing used here is a continuous strip of metal flashing at least 10 inches wide. Use a long straightedged board to bend the flashing in the middle to match the angle of the roof and wall.

If the vertical wall has not yet been covered with siding, the job is quite simple. Slip the flashing under the felt on the vertical wall and embed the flashing on the roof in a layer of roofing cement. Nail the flashing to the roof but *not* to the vertical wall. This allows the house to settle at a rate different from that of the attached roof without disturbing the flashing seal.

For aluminum flashing, use aluminum nails, and use galvanized nails on galvanized flashing.

With the flashing in place and embedded in cement, apply the final course of roofing material and cover each nail head with a dab of cement. When you apply the siding on the vertical wall, *do not* nail through the flashing.

Where the vertical siding is already in place, the flashing must be slipped under it. Gently pry the siding away from the wall and work the flashing up under it. If you run into nails, notch the flashing to fit around them.

If both the vertical wall and the roof are covered with wood shingles or shakes, the final course on the roof may be too short for a neat appearance. Instead of using short shakes or shingles, cut a bevel on a length of 1-by-4 redwood or cedar and install as illustrated. The bevel cut will ensure a snug fit under the last course of shingles on the wall.

If the siding is stucco (or brick), the flashing must be set into the concrete. The first step is to bend the top ½ inch of the continuous flashing at a 90-degree angle. Do this by clamping the flashing between two boards placed in a vise, and bending it.

Snap a chalkline across the stucco wall about 5 inches up from the roof and use a cold chisel to remove the stucco at least ½ inch deep along the line. On a brick wall, chisel out the mortar between the bricks about 5 inches up from the roof. Bend the top edge of continuous flashing to fit into the groove; bend the middle so that the lower half will fit smoothly on the roof. Fill the groove with mortar or caulking and press the top edge into place. Nail the last course of the roofing material over the flashing on the roof.

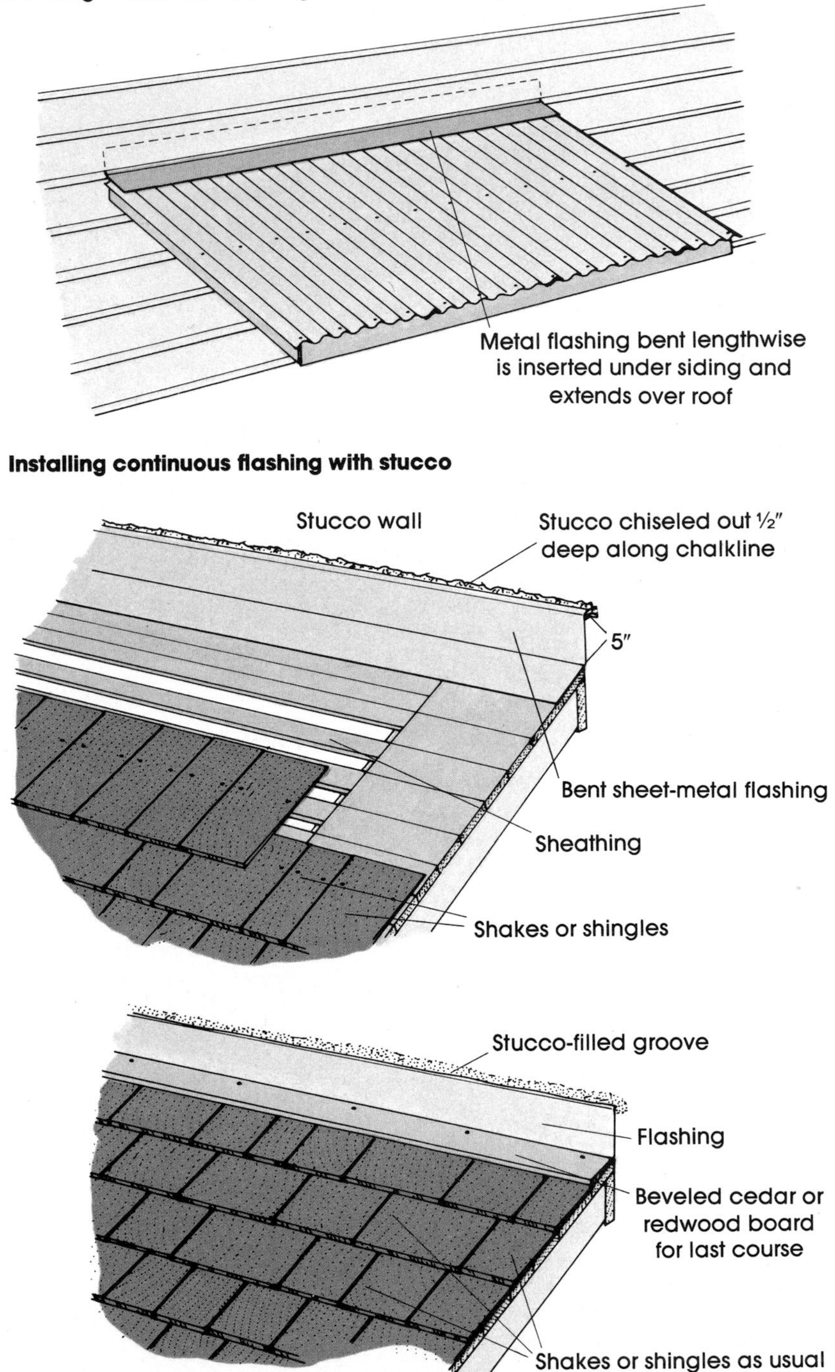

Step Flashing

Step flashing is applied where the sides of a dormer meet the roof, and is also part of chimney flashing. Each course is protected by its own flashing "shingle" cut from aluminum or galvanized tin, 10 inches long and 2 inches wider than the exposure. On an asphalt roof with a 5-inch exposure, for instance, the flashing shingles are 7 inches wide.

You can make your own step shingles or have them made at a sheet-metal shop. To bend your own, cut the strip, then clamp it between two boards in a vise and hammer it over with a rubber mallet, as shown, so that 5 inches will extend up the vertical wall and 5 inches will extend over the roof. The roofing felt should extend up the vertical wall about 5 inches.

The step flashing process described below uses asphalt shingles with a 5-inch exposure as an example. The process is the same for wood shingles, shakes, or tile. (For roll-roofing and panel roofs, use continuous flashing.)

To apply the flashing, put the first piece on top of the starter shingle and position it so the bottom edge of the flashing is flush with the bottom edge of the starter course. Nail the flashing to the roof with two nails placed 1 inch from the top. Do not nail any step flashing to the vertical wall.

If the siding is already in place on your house, you will have to pry it away and slip the flashing up under it. Cut notches in the flashing to fit it past siding nails. For brick or stucco houses, cap flashing must be applied over the step flashing as described in the following section on flashing a chimney.

After the first piece of flashing has been applied over the starter course, apply the first shingle course with its butt end flush with the lower end of the starter course. Measure 5 inches up from the butt of the first shingle and apply the next piece of step flashing with the bottom end along this line. Nail the top, then put the second course of shingles over it with a 5-inch exposure. Continue up the roof.

Bending step shingles

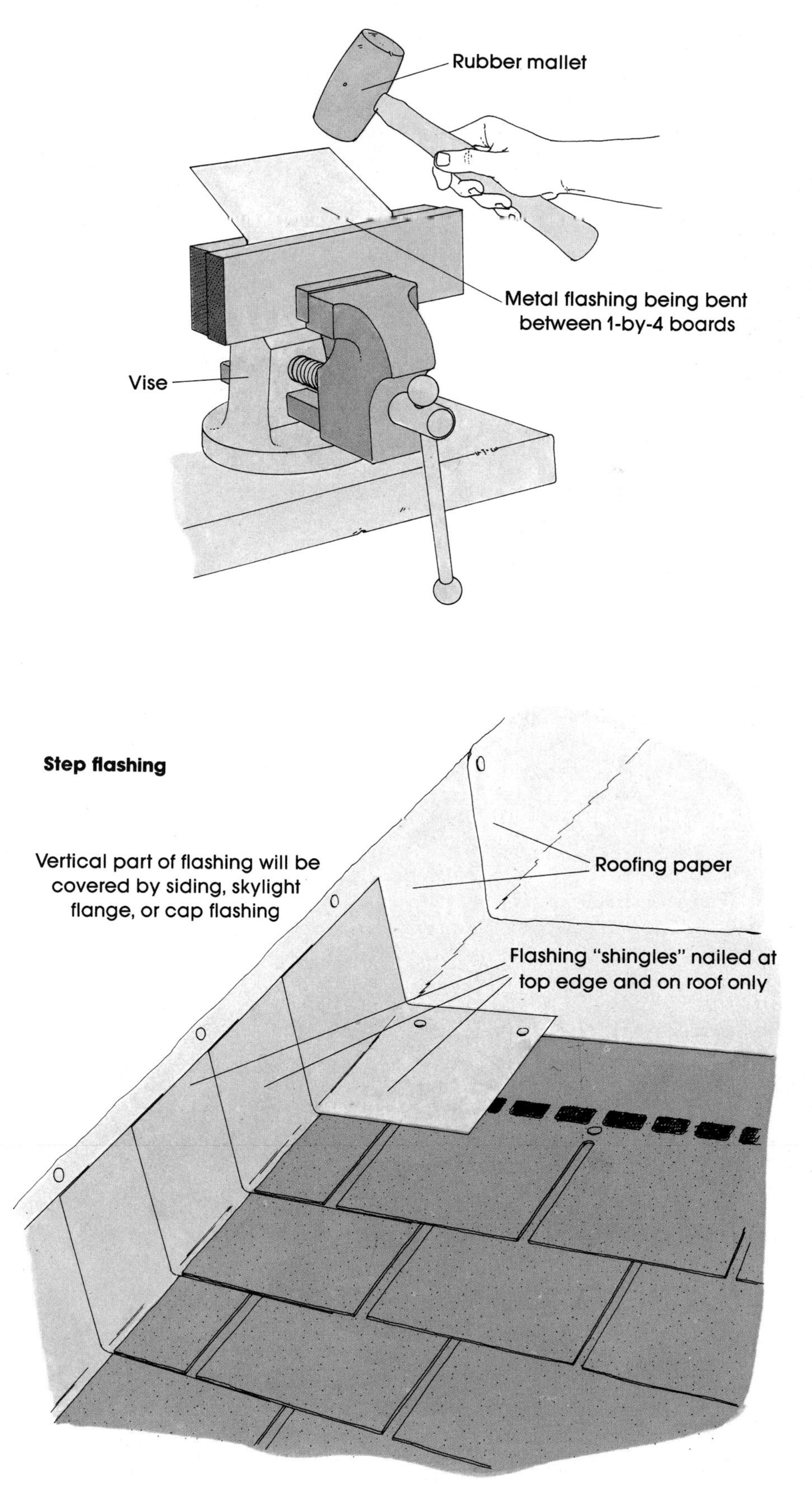

Step flashing

Flashing a Chimney

This is the most difficult and important aspect of flashing a roof, since many leaks originate around chimneys. Although a careful nonprofessional can do it, call in a contractor if you have doubts.

Chimney flashing consists of aluminum base flashing and cap flashing (also called counter flashing). The two overlap but must not be joined together, since the chimney and house settle at different rates. Start flashing the chimney after you have installed your roof up to the chimney base.

If your chimney is wider than 2 feet, or if you live in an area of heavy snow and ice, you will first need to construct a "cricket" along the up side of the chimney. Cut it from two pieces of ¼-inch exterior plywood (see illustration) and apply it to the roof deck. The cricket diverts snow and water that would otherwise build up behind the chimney and possibly cause leaks.

The next step is to coat the bricks around the chimney base with asphalt sealant. As each piece of base flashing is installed, it will be pressed into roofing cement spread on the bricks. Sealant makes the cement adhere properly to the bricks.

Now cut a piece of base flashing for the down side of the chimney, as illustrated, and bend it between two boards in a vise to fit around the chimney. Embed the apron in cement on the roof and press the flanges into cement spread on the bricks. You can hold the flanges in place by driving a couple of nails through them into the mortar. There's no need to remove the nails.

Continue roofing alongside the chimney, applying step flashing as described on page 57. Each piece of step flashing must be embedded in cement against the chimney, and the end of each shingle placed on the flashing must be embedded in cement. Note how the first and last step flashings are cut and bent to fit around the chimney.

Now cut and bend a piece of base flashing to fit around the chimney and over the cricket, and at least 6 inches up the side of the chimney, as illustrated. In addition, the base flashing should extend beyond the cricket and onto the roof by 6 inches. Nail it to the roof. This is not necessary, however, if the cricket is large enough—2 feet or longer—to be shingled (in such a case, shingle it as you would a dormer roof; see page 42).

Now comes the installation of the cap flashing. It should be set in the mortar course two bricks above where the base flashing reaches up the chimney, and extend down to within 1 inch of the roof. Use a narrow cape chisel to remove the mortar between the bricks to a depth of 1½ inches.

The first piece of cap flashing is installed on the down side of the chimney. Cut and bend it, and fit it into the gap you've chiseled out. Pack the joint with Portland cement mortar (see box).

Cut enough pieces of cap flashing to cover the two sides. Note how the last piece is bent to fit around the corner of the chimney on the up side. Each piece should overlap the next by 3 inches. Finally, cut and fit the cap flashing on the up side of the chimney, working around the cricket.

Pattern for lower base flashing

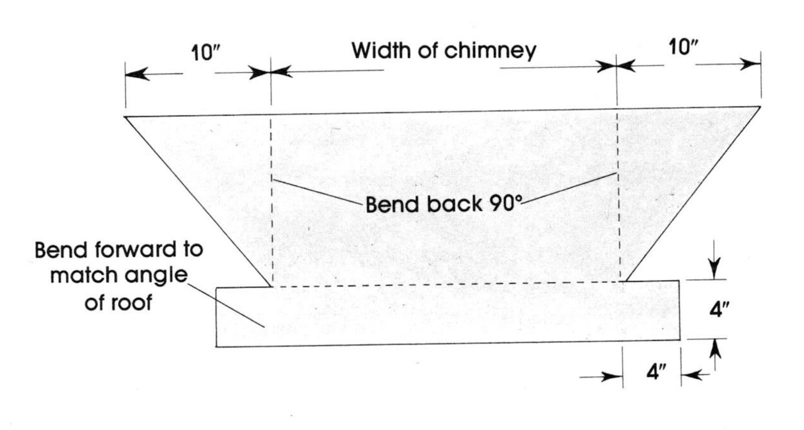

Applying step flashing

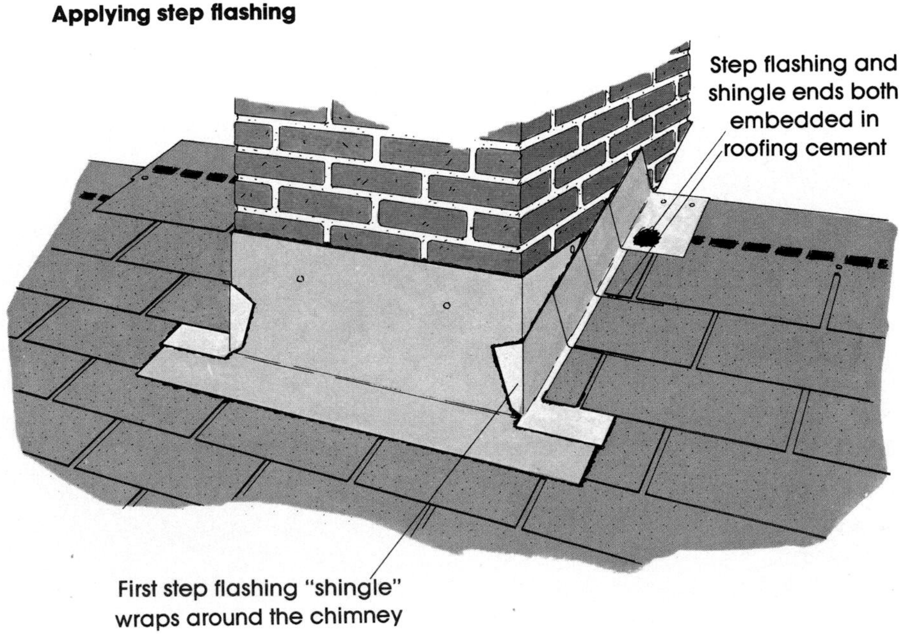

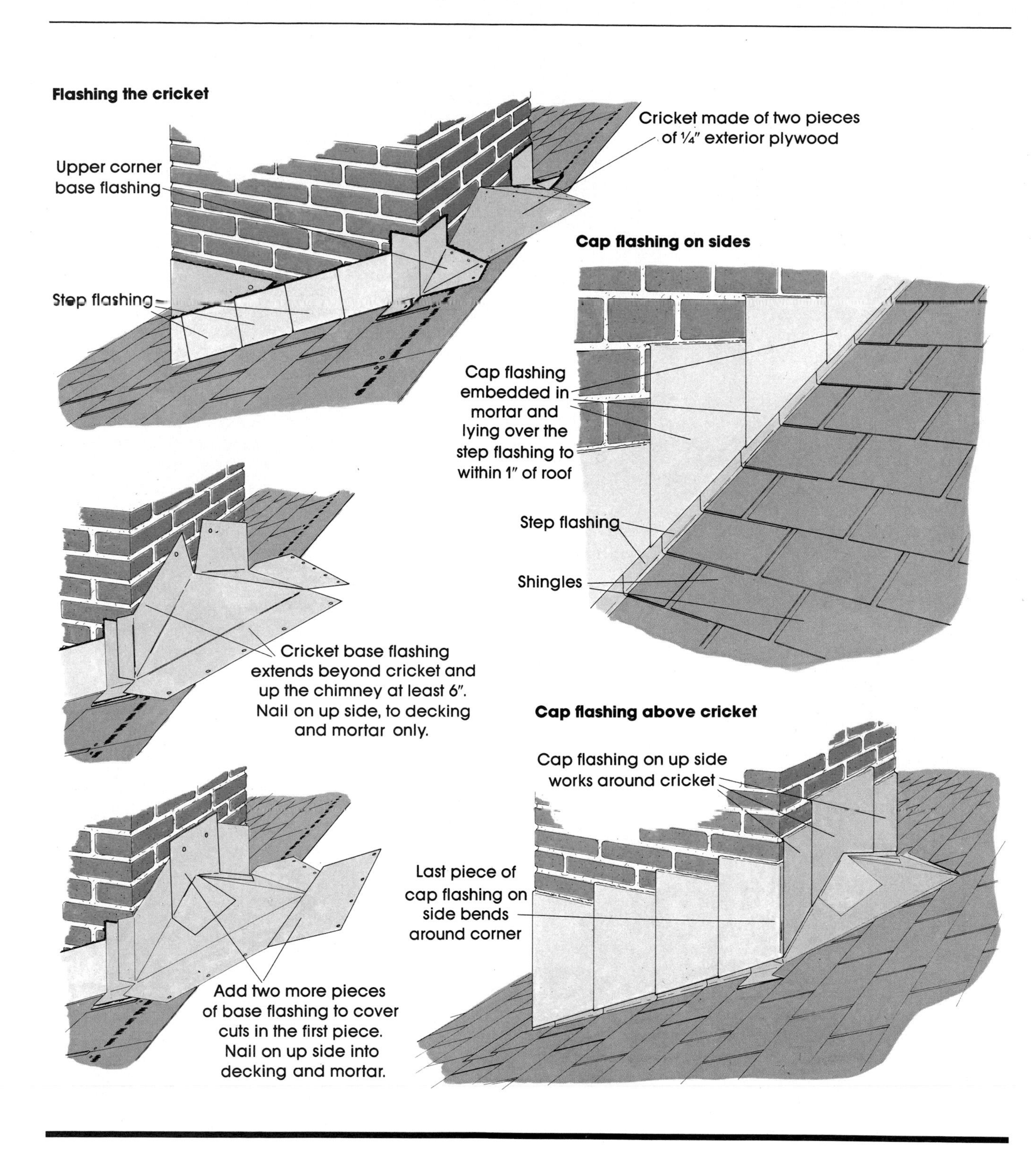

Mixing and Applying Mortar

Cap flashing in a chimney must be held in place with mortar. Make the mix by combining 1 part Portland mortar cement to 3 parts of *fine* mortar sand. Three cups of mortar and 9 cups of sand will probably be enough. Put the mix in a bucket, then slowly add water and stir constantly. Add water just until it is the consistency of thick whipped cream. The sand particles should be mixed so thoroughly that they are not visible.

Use a wire brush to scrub out the joints where you removed mortar. Wet the joints thoroughly, then apply fresh mortar.

Fill the joint with mortar, then insert the flashing strip in the center of the opening. Push it firmly into place. Smooth and press the mortar above the flashing with your finger.

After a couple of hours, wet the mortar with fine spray, then wet it again the following day if you remember. This will help it set slowly and keep it from cracking.

INSTALLING

GUTTERS & DOWNSPOUTS

Gutters are installed on eaves to collect runoff from the roof and channel it into downspouts that direct it away from the house foundation. Without gutters and downspouts, water can drip from the eaves onto people's heads and onto border gardens, causing bad tempers and soil erosion. Worse, water can saturate the soil, then seep into the crawl space or through basement walls.

There are different styles of gutters and downspouts, but they connect in a similar fashion and are mounted along eaves in the same manner.

Selecting Gutters and Downspouts

Gutters and downspouts are most commonly made from galvanized metal, aluminum, or vinyl. Copper and wood gutters exist, but they are more expensive. Aluminum and galvanized metal are sold either factory-painted or bare, which allows you to paint them to match your house, if you wish. Vinyl comes in a wide variety of colors.

Gutters and downspouts are normally sold in 10-foot lengths, which are the easiest to handle; longer ones can be specially ordered. Gutter widths are usually 4 inches, 5 inches, or 6 inches.

As a general rule, roofs of 750 square feet or less use 4-inch-wide gutters; roofs up to 1500 square feet use 5-inch gutters; and roofs with more than 1500 square feet take 6-inch gutters.

Gutters are sometimes half-round but more commonly have what is called the forged shape. Downspouts are either round or square, and are often corrugated, giving them additional strength.

The gutter sections can be hung from the eaves with any of a variety of devices. The spike and ferrule, or the two different styles of brackets, are quite commonly used. Strap hangers provide more support, but must be installed before the roof is put on. It is virtually impossible to nail them to a completed roof without damaging the roofing material. Downspouts are secured to the side of the house with galvanized straps.

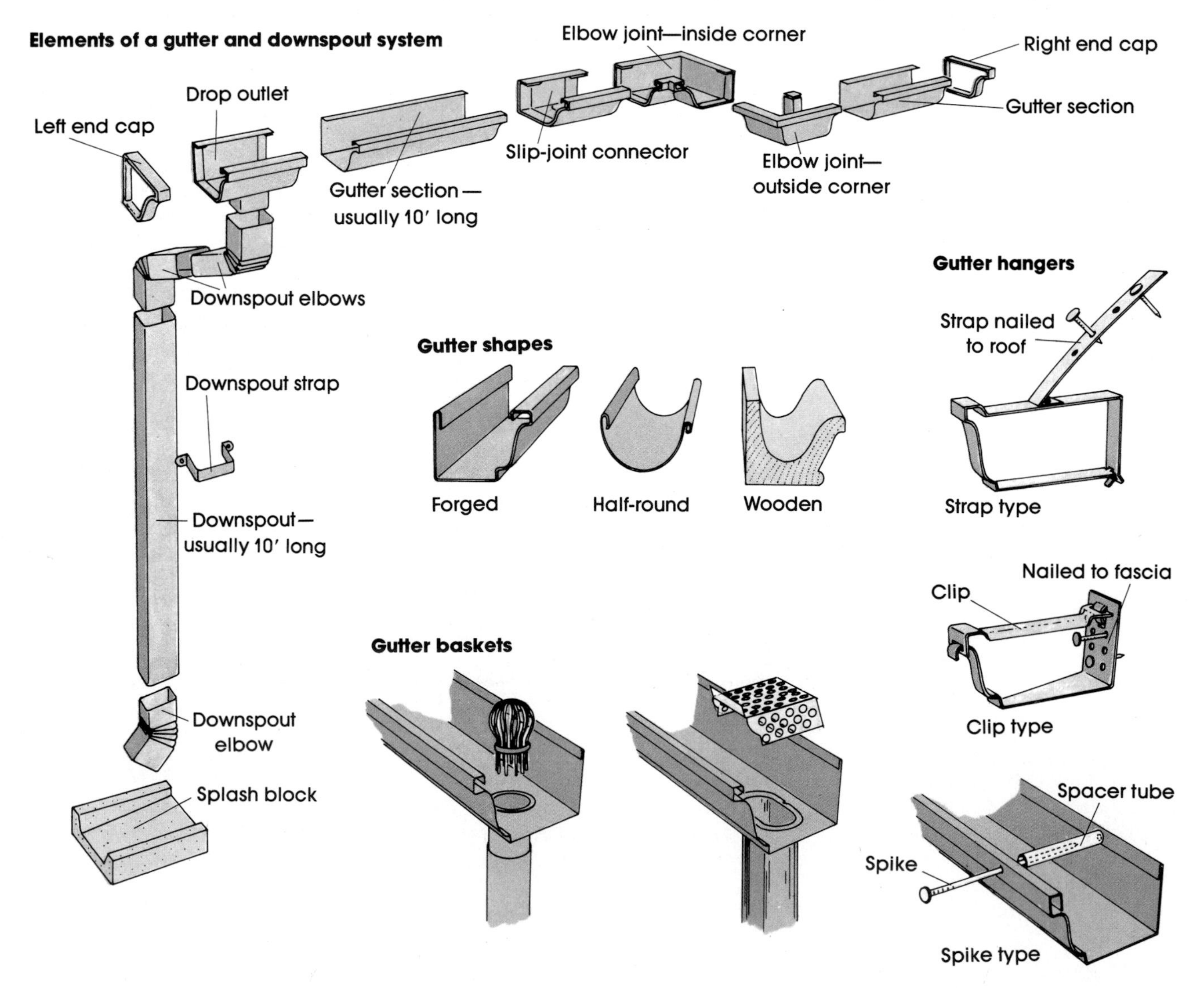

Estimating Your Needs

First measure the length of all the eaves to calculate the number of gutter sections and supports—one every 3 feet to 4 feet—you will need. Count the number of inside and outside corners. Figure the number of left and right end caps.

A drop outlet is needed for every 40 feet of gutter, so calculate the number needed. Three elbows are needed for each drop outlet: two to reach the side of the house and one on the end of the downspout pipe. Count the number of downspouts needed and add a few extra if some must be cut and used as connectors between the elbows at the top. Don't forget the straps for the downspout pipes, one for every 6 feet of pipe.

Now count how many slip connectors you need to join the pipe, and remember that you do not need one where the gutter sections meet at corners or drop outlets.

Finally, add up the number of splash blocks or leaders you'll need under the downspouts.

Installing Gutters and Downspouts

Gutters should slope about 1 inch for every 20 feet. If you have a run of 40 feet or more, then slope the gutters from the middle of the run and put a downspout at each end.

To lay out the gutter slope, tack a nail to the fascia board at the high end of the slope, measure the run, and drop 1 inch every 20 feet. Tack a nail at this position on the other end and snap a chalkline between the two points. Use the line as a guide.

Lay out all the components on the ground below the eaves. Measure the gutter runs and note the downspout locations, then cut the gutters accordingly. If gutters are unpainted metal or plastic, cut them with a hacksaw. Use tin snips on painted gutters to minimize the shattering of enamel paint. To steady the gutters while sawing, slip a length of 2 by 4 in the gutter about 1 inch back from the cut, then squeeze the gutter against the block. Use a file to remove all burrs (jagged edges) from the cut edge.

Gutters should be installed by two people, if possible. One supports the far end while the other installs the gutter and its hangers. If you don't have a helper, hang the far end in a loop of string from the guide nail, then work toward it.

When all pieces are connected and secure, go back and seal each joint with caulking to prevent leaks.

Connect the downspout elbows to the drainpipe on the drop outlet by drilling holes on opposite sides and inserting sheet-metal screws. Connect the elbows to the downspouts in the same manner.

Bend the straps to fit the downspout, then screw to the siding (the method is the same for all materials). Fit the elbow on the end of the downspout and put the splash block under it. If you wish to carry the water farther from the house, attach a length of downspout to the elbow. This extension can be buried and run to a drywell, as described below.

Finally, put strainer baskets over the downspout holes and aluminum or vinyl mesh over the gutters. They'll keep out leaves and debris, which can produce clogs. These two items are available wherever gutters are sold.

Installing Drywells

If you have trouble keeping roof runoff water diverted from your foundation, you can install drywells, one for each downspout. These are holes about 4 feet across and 4 feet deep placed 6 feet to 10 feet from the house. Runoff water from the roof is directed into the drywell, where it will seep deep into the ground.

Bury a length of downspout pipe from the downspout to the well with just enough slope to ensure good water movement. Fill the hole with coarse gravel and replace the sod on top.

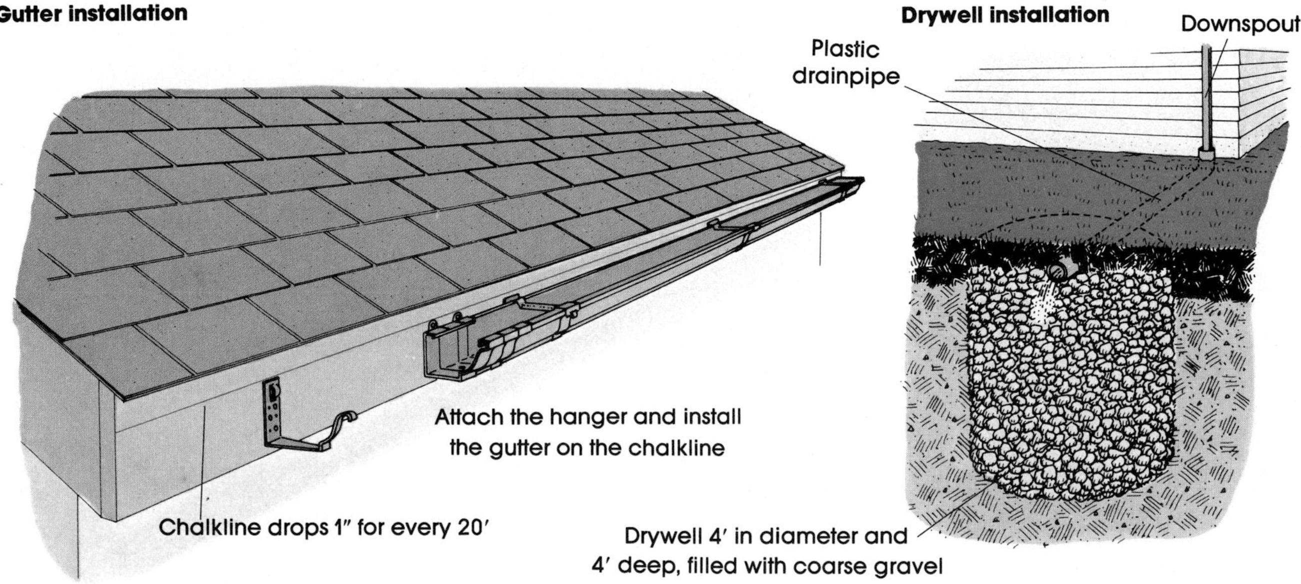

SIDINGS

Applying siding is a good do-it-yourself project.
This chapter contains complete instructions
for installing horizontal and vertical
board siding, plywood and hardboard panels,
wood shingles, and stucco, with
directions on how to select the siding,
prepare the wall, and cover gable ends and corners.

Whether you are working on new construction or planning a major facelift for your existing home, choosing the siding should be done with all the care and consideration you give to the home's interior. The guide to sidings on pages 64–65 will help you consider your choices. There are a number of aspects to the decision, among them your own taste, the style of your house, the prevailing styles in the neighborhood, quality, cost, and the complexity of the installation process.

First, the siding should suit your taste. It is going to say something about your style of living. Do you want the rustic effect of shingles or board-and-batten; the sophisticated simplicity of narrow vertical strips of redwood or cedar; practical stucco; or the clean horizontal lines of traditional shiplap?

The style of your house will dictate some choices in both materials and colors—a saltbox seems to demand shiplap, a ranch house may call for vertical siding or panels, a restrained Georgian facade shouldn't be covered in natural wood shakes. The roof may also suggest a particular siding. One often sees cedar or redwood siding with a shake roof, but going against tradition—say, stucco siding and a shake roof—can also produce pleasing results.

Any choice you make should take into consideration the prevailing styles and color combinations in the neighborhood—you don't want your house to stand out in an unneighborly way.

Applying your own siding is a good do-it-yourself project, and you can save 50 percent or more of the overall cost by not hiring a contractor. Other factors influence the cost, however. Do you want the low maintenance of aluminum, vinyl, steel, or brick at the additional price of having it professionally installed because of its complexity? Stucco (see pages 78–83) is a high-quality, low-maintenance material that can be applied by the do-it-yourselfer—if you are very patient and painstaking. Shingle siding requires more time but is easily installed by one person (see pages 76–77). The same is true of the individual boards used in horizontal and vertical siding (see pages 69–73). They are also less expensive than wood shingles—but the shingles have a lot of visual appeal, and they require less maintenance. Plywood panels can't really compete with some of the other materials for beauty. But they are among the least expensive and, if you have at least one helper, they go up fast.

Whichever siding you finally choose, it will not be inexpensive; but good new siding, carefully applied, will enhance the appearance and value of your house. You'll be glad you spent the time and money when the first winter storm hammers those weathertight, secure walls.

The rich glow of newly applied cedar-shingle siding warms this home, also seen on page 1. Together with the corrugated-metal roofing, it creates a durable and colorful exterior.

Installing New Siding Over Old

In a re-siding project, new siding can often be applied over the old; but if your house is not insulated, you may choose to rip off the old. With the siding removed, you can install fiberglass insulation batts. In addition to cutting your fuel bills, the insulation—always placed with the vapor barrier toward the warm side of the house—will prevent moisture inside the house from moving through the walls and literally forcing the paint off the exterior walls of your home.

If you decide to put the new siding directly over the old, you will eliminate a big removal job, and your house will stay protected while you work at your own pace. You can still install insulation if your house does not already have it. Sheets of rigid insulation, called beadboard, can be nailed to the existing siding and the new siding placed over it. This procedure, in addition to insulating the house, also provides a smooth surface for attaching the new siding.

A CHOICE OF SIDINGS

Choosing the siding for your house is akin to choosing new furniture or a major appliance; it must be durable, attractive, and in your price range. When you combine those prerequisites with the wide array of siding available, making the choice can be difficult. Here is a brief survey of the possibilities.

Wood

Wood siding, in all its varied forms, still remains the most widely used residential siding in the country. It also offers a confusingly broad variety of choices.

Panels. Wood panel sidings are made of plywood or hardboard, which is constructed of heat-processed wood pulp pressed into sheets. Panels are normally 4 by 8, 4 by 9, and 4 by 10. Plywood siding styles include smooth and rough finishes, and grooved panels that imitate board siding. Hardboard siding comes in an even wider variety of styles, ranging from stucco to embossed sheets that resemble shingles.

Panel siding has the advantages of relative low cost and ease of application. Two experienced workers can cover an average house in a weekend or less. Redwood and cedar panels are also low-maintenance—they can be allowed to weather naturally—but other types of wood panels must be stained or painted regularly.

Boards. Solid-wood siding, both horizontal and vertical, runs the gamut of styles, some of them illustrated here.

Shingles. Shingle siding, which gives a rustic look, is another top choice in wood. The relatively high cost of

Two plywood panel siding styles

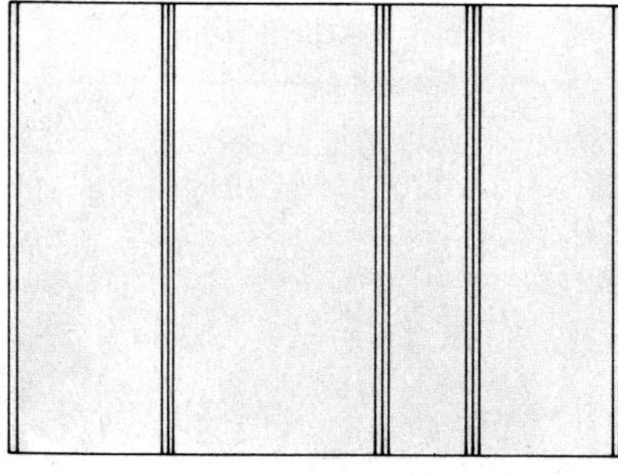

Random board V-groove, smooth finish

Batten-and-board grooved, resawn finish

Board siding

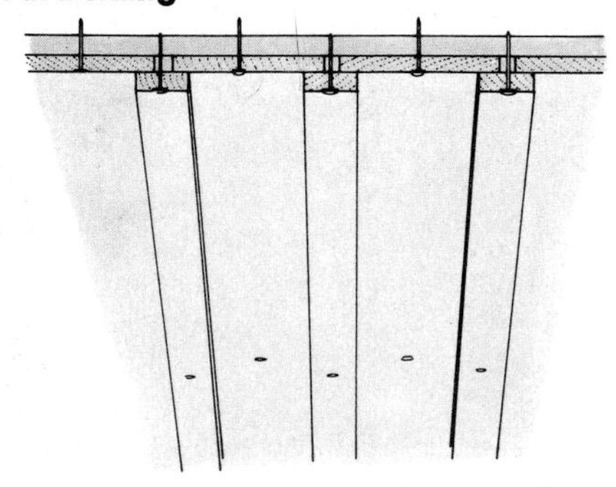

Vertical board-and-batten

Board siding

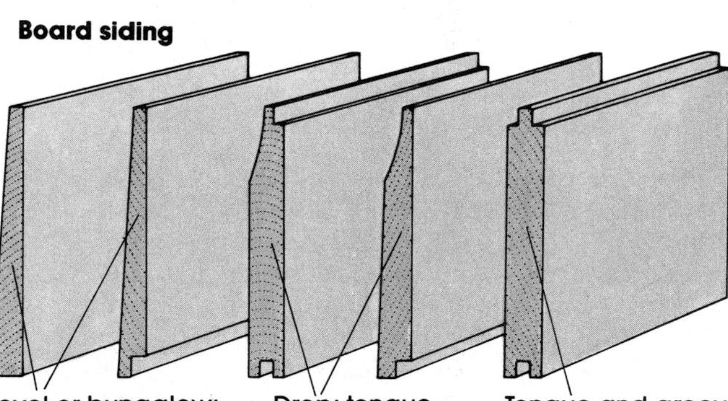

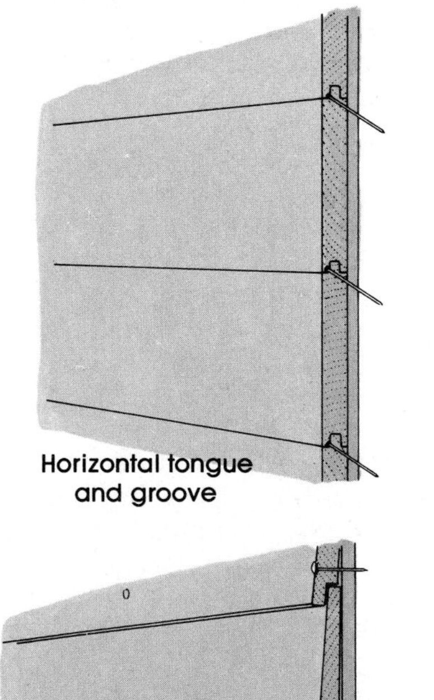

Horizontal tongue and groove

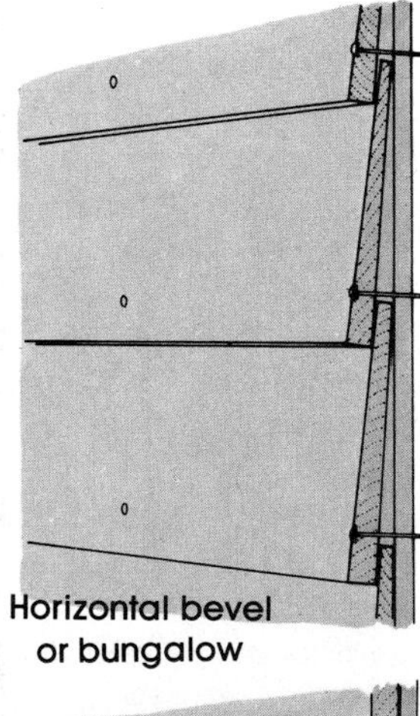

Horizontal bevel or bungalow

Horizontal Dolly Varden

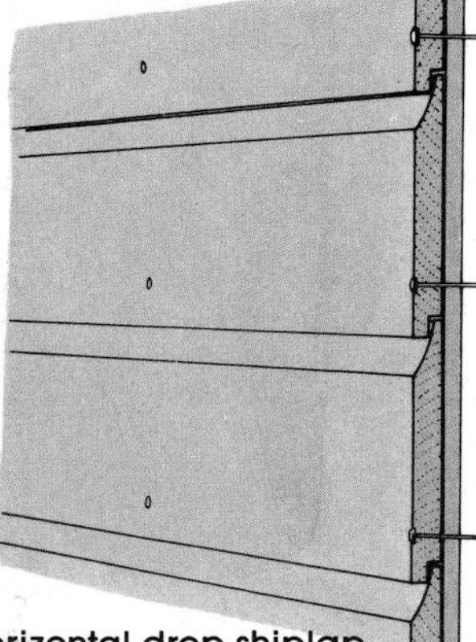

Horizontal drop shiplap

Shingle patterns using standard shingles

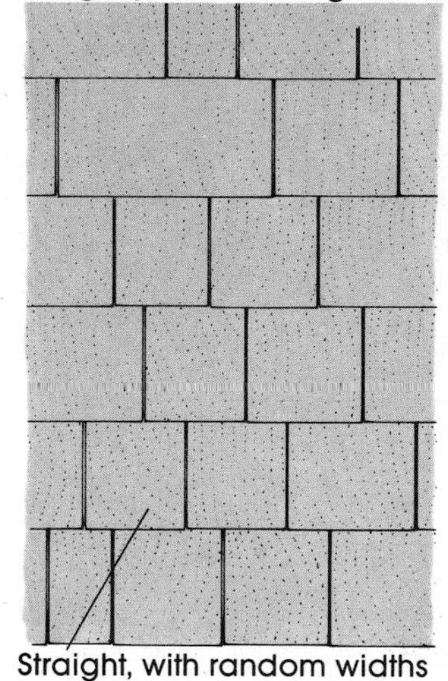

Straight, with random widths

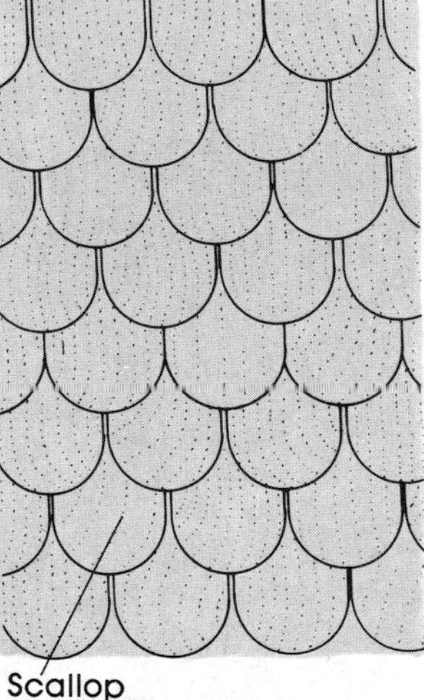

Staggered, all the same width

Shingle patterns using decoratively cut shingles

Scallop

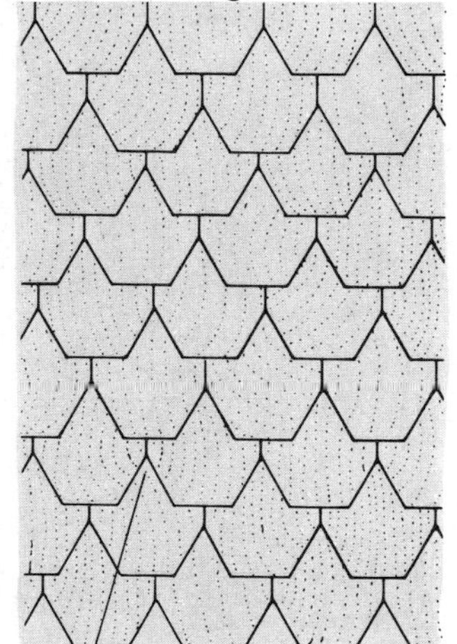

Angled scallop

Stucco patterns

Modern American

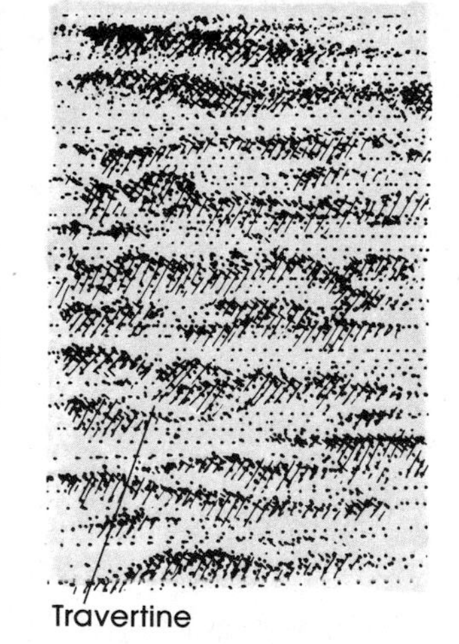

Travertine

Spatter

Old English

shingles is offset by several factors: you can install it alone; it needs no painting; it will last for years.

Stucco

Stucco is one of the most durable sidings available for houses. Made of concrete, it is applied in three separate coats, with the desired color pigment mixed into the finish coat, so no painting is required. Applying it yourself will cut costs by 50 percent or more. Stucco does present some problems, though. It is physically difficult to apply, requires careful application, and can crack if applied incorrectly or if the house settles unduly.

Metal and Vinyl

Three other types of siding that are rapidly gaining in popularity deserve a mention here: aluminum, vinyl, and steel. This book does not cover their installation, which requires professional skills, experience, and special tools. However, you may want to read through the information that follows and consult a contractor. If you wish to attempt installing one of these sidings yourself, contact a manufacturer and request a copy of the installation guidebook.

Aluminum. Aluminum siding comes in two basic styles, to give the appearance of either wide or narrow horizontal board siding, and in a choice of several colors of factory-applied enamel coatings. The chief advantages of aluminum siding are its long life and its relatively low maintenance. Aluminum itself is a poor insulator, so it is often applied over sheets of rigid insulation, which also provides soundproofing for those who live in areas of frequent hail storms. Aluminum siding can be dented easily, which may cause the enamel coating to break. However, individual panels can be replaced.

Vinyl. Vinyl is applied in much the same way as aluminum and has a similar appearance when completed. Being a plastic, vinyl is more flexible and easier to work with than aluminum, but there is still a great amount of precise cutting and fitting. Vinyl does not dent like aluminum and chips are not a problem because the color is uniform throughout. However, vinyl expands and contracts more than other materials, and this must be allowed for during installation. One drawback to vinyl is its brittleness when subjected to extreme cold. Cracked or broken panels can be easily replaced, though.

Steel. Steel siding continues to grow in popularity as a residential siding. Like aluminum and vinyl, it comes in different colors and is extremely durable. However, it definitely requires professional installation.

PREPARING TO INSTALL SIDING

Jamb Extenders

When new siding is applied over old, the added thickness often means that the window jambs no longer extend beyond the siding. This is corrected by applying jamb extenders around the doors and windows before re-siding.

Jamb extenders are made by ripping jamb stock to the same width as the thickness of the new siding. The stock must be cut accurately and smoothly; use a table saw, a radial arm saw, or a circular saw guided by a cutting jig.

Installing Jamb Extenders

Wood windows. For either a door or a window, carefully pry off the exterior casing (trim) and set aside for reuse. Nail the top jamb extender to the edge of the top jamb, then butt the two side extenders against the top piece. Trim the bottoms of the side extenders to match the angle of the sill, then nail in place. Set the nails with a nail set and fill the holes with wood putty.

After the new siding has been installed, cut a length of continuous aluminum flashing the length of the casing above the door or window. Bend it to fit over the casing by placing it between two boards in a vise. The flashing should slip behind the siding above the window casing and fit over the top of the casing. If this is not practical for the type of siding you are using—panel siding, for instance—lay a bead of caulking in the angle where the casing meets the siding.

When the flashing is on, put the casing back in place as before. Set the nails, cover with wood putty, and paint.

When window jambs must be extended, often the window sills must also be extended. If the sill is round-nosed, plane it flat. Rip a length of sill extender from stock the same thickness as the existing sill and nail it in place with casing nails. Set the nails and fill the holes with wood putty. Fill any uneven spots in the joint between the extender and sill with wood putty and sand when dry. Finally, repaint the jamb extenders and sill.

Metal-framed window with jamb extenders

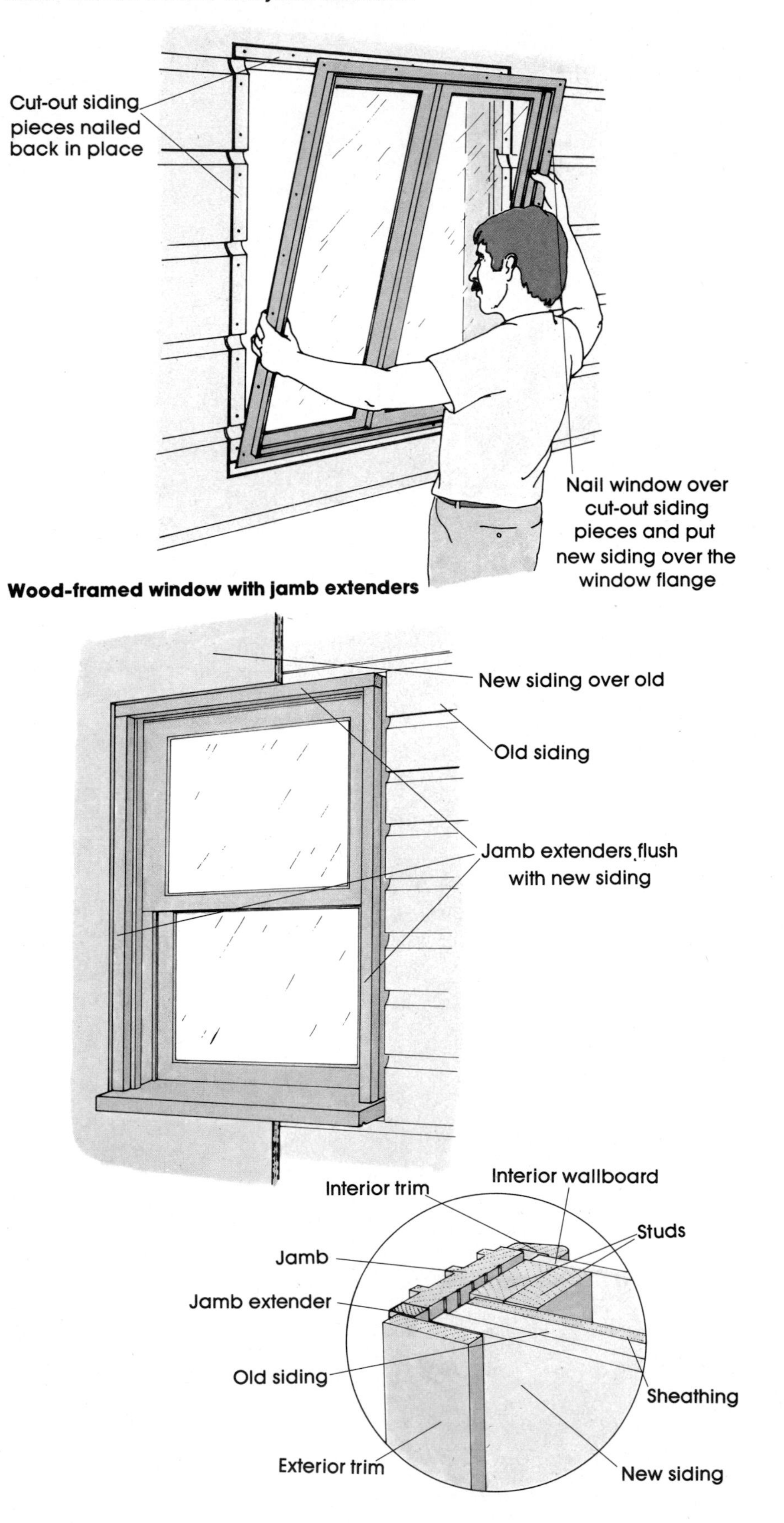

Wood-framed window with jamb extenders

Metal windows. Since jamb extenders obviously cannot be nailed to metal windows, the windows must first be removed. First remove the exterior window casing. Metal windows are held in place with nails driven through flanges around the windows. To get at those flanges, you must remove the siding next to the window. The flanges are about 1 inch wide, so measure 1½ inches out from all sides and snap a chalkline. Set the blade on your circular saw ⅛ inch deeper than the thickness of the siding and cut along the lines. Save the cut-off pieces of siding. Use a straight claw hammer to remove the nails in the flange; remove the window.

Now nail the cut-off pieces of siding back in place around the rough opening (you may need to predrill holes to avoid splitting the siding). Place the window in the opening, check that it is level, and nail it in place through the nailing flanges.The new siding will be brought to the metal edge of the window and over the flange, and then the casing will be put back to cover the gap between the new siding and the window.

The Story Pole

The story pole is used with shingle and horizontal board siding to keep courses straight and level, and to adjust the siding exposure widths so there is a minimum amount of cutting above and below window openings. The least cutting and fitting occurs when window bottoms are all the same distance from the ground.

The story pole should be a straight 1 by 2 or 1 by 4 that is long enough to reach from a spot 1 inch or more below the existing siding, or below the top of the foundation, to the top of the wall.

To lay out the story pole, place it against the side of the house next to a window. The top must be flush with the top of the wall and the bottom must extend below the sheathing. Mark the story pole at the bottom of the sheathing, the bottom of the window sill, and the top of the window drip cap.

Lay the pole down. With a pair of wing dividers set at the desired siding exposure, mark off the distance between the window sill and the drip cap. If it comes out even, you are in luck. If not, adjust the wing dividers until they mark off equal portions. The length of your divisions must be less than the maximum exposure recommended for the siding. Your supplier will provide specifics on exposure, but as a general rule, a clapboard siding board should overlap the one below by 1½ inches, and shingles should have an exposure of ½ inch less than half the overall length of the shingle.

By ensuring in advance that the butts of a shingle course, or the top edge of a horizontal siding course, line up with the bottom of the sill and the top of the drip cap, you will make your siding look neater and you will have less trimming to do around the windows.

Once you have established the proper intervals, mark them on the rest of the pole and use a square to make the lines straight across the story pole.

Place the pole at one corner of the wall to be covered and align the bottom mark with the sheathing edge. Transfer the story pole marks to the corner boards or sheathing. Do the same along remaining corners, doors, and windows. As you begin installing siding, use these marks to set up your guide—either a chalkline or a straight board.

Making a story pole

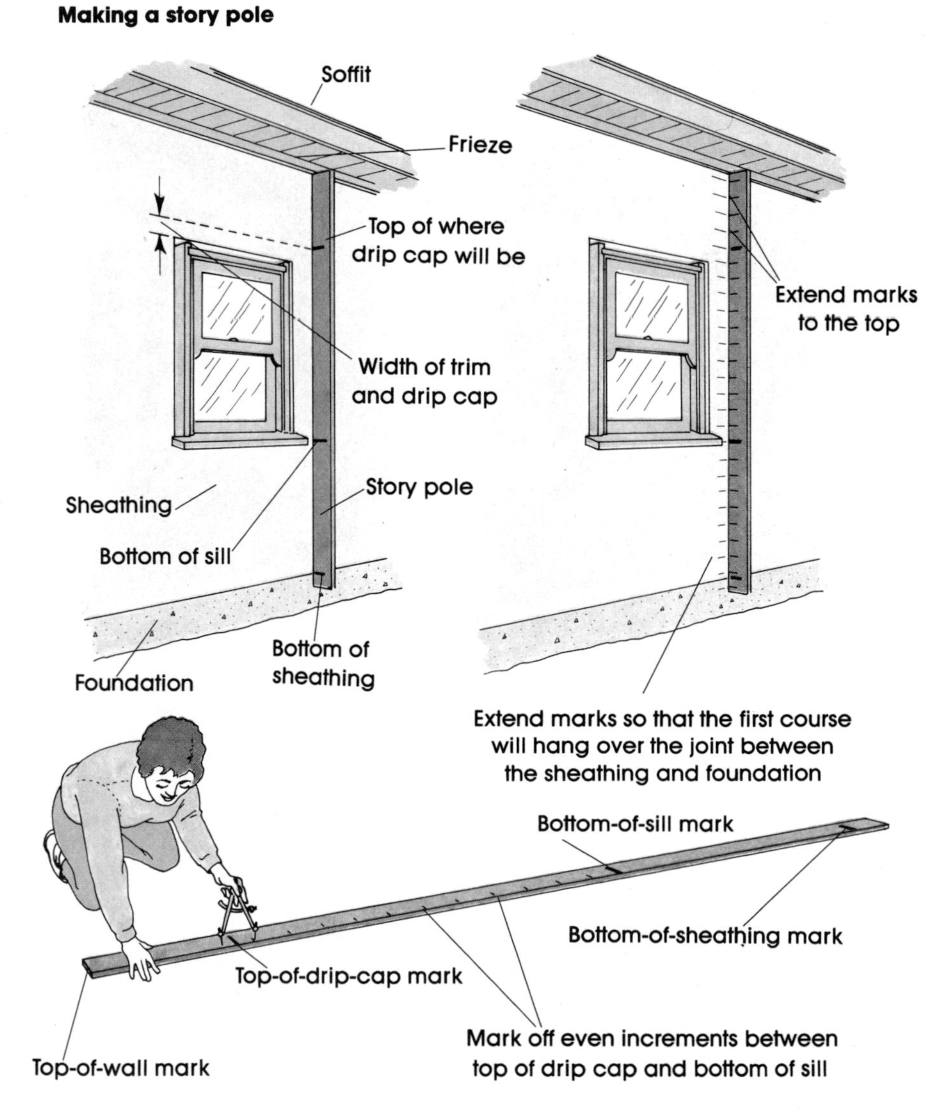

PREPARING TO INSTALL SIDING

CONTINUED

Bracing the Wall

The exterior walls of a house must always be braced to prevent lateral shifting. The method you use will depend in part on what type of siding you use.

If your siding consists of plywood or hardboard siding, the large panels themselves provide all the lateral bracing required by most building codes. The ⅜-inch plywood sheathing applied over the studs preparatory to putting on stucco siding is also sufficient. But if you are applying horizontal or vertical siding (whether board, metal, or vinyl), or shingles, you must use let-in braces or metal straps.

Let-in Braces

Let-in braces are 1 by 4s set into the edges of the studs forming the walls. They must always run from the wall's outside top corner to its bottom center. Place the 1 by 4 across the wall from the cap plate to the sole plate at approximately a 45-degree angle and mark the studs where the 1 by 4 crosses them. Cut notches in the studs with a circular saw and finish them with a chisel. Nail the 1 by 4 to the notches and the plates. For more details on installing let-in bracing, see Ortho's book, *Basic Carpentry Techniques*.

Metal Straps

Metal straps are 12 feet to 16 feet long, with nail holes about 1 inch apart. They are best placed at the corners of the building. They must reach from the cap plate to the sole plate, pulled tight, and must be nailed with 16d nails to every stud they cross. Most importantly, metal bracing straps must always be used in pairs so they cross at the center of the wall to form an X.

Flashing

Flashing is needed above wood-framed doors and windows; it is not necessary with metal-framed windows. Lightweight aluminum is most widely used. It is commonly sold in 8-inch-wide rolls in lengths of 10 feet or more.

To fit the flashing above a door or window casing, first measure the thickness of the casing. Add ½ inch to this measurement and snap a chalkline at that point on the flashing. Place the flashing between two boards in a vise and bend it along the chalkline to a 90-degree angle. Put the flashing in place above the door or window. The ½ inch extra will protrude over the casing. Tack the upper edge of the flashing to the sheathing or header with roofing or shingle nails.

To bend the ½-inch protrusion down over the casing, cover it with a length of 2 by 4 and hammer it down flush with the casing.

When installing shingles over the flashing, keep the butts at least ¼ inch away so they will not soak up any collected water.

Building Paper

In most cases, you will need to cover the wall with building paper before installing siding (see the instructions for particular sidings, as requirements vary). It should overlap the flashing to about an inch from the door or window casing.

Let-in bracing

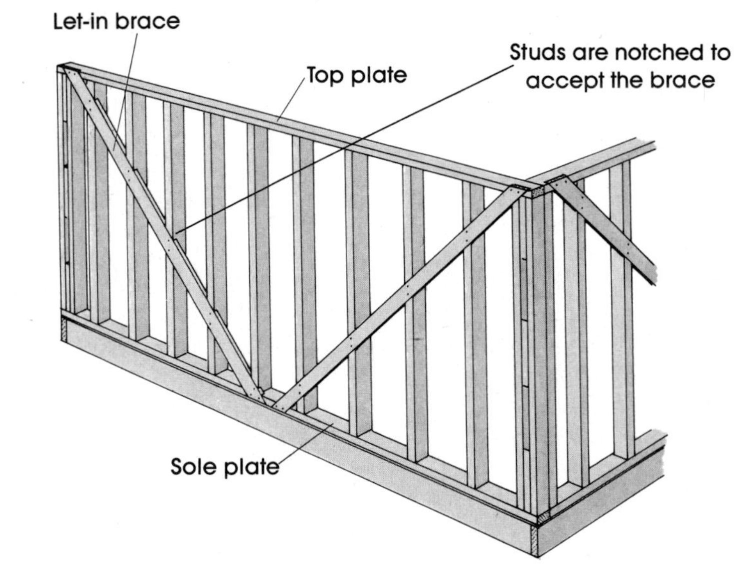

Metal strap bracing

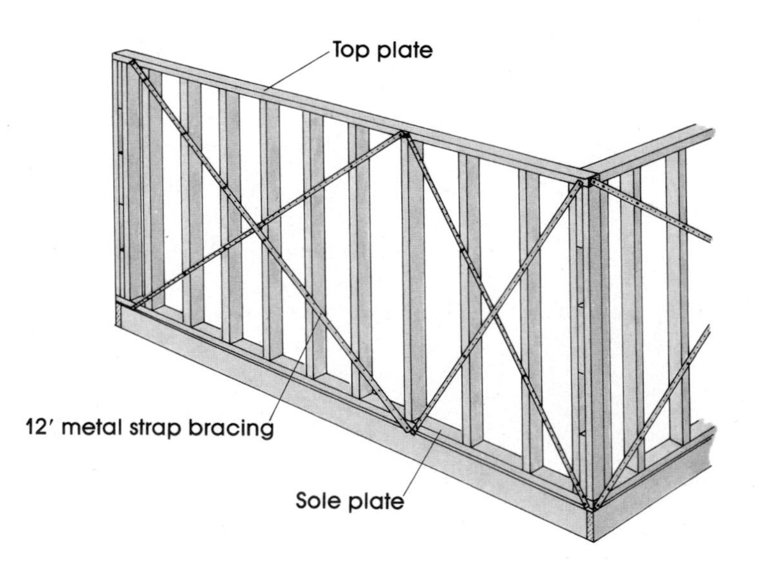

HORIZONTAL WOOD SIDING

Wood siding, whether the classic shiplap of a New England saltbox or the vertical redwood siding of a Pacific seacoast house, remains a favorite among homeowners. The long clean lines will enhance the appearance of virtually any house. This type of siding may not go up as fast as panel siding, but the individual boards are much easier to handle.

Redwood and cedar siding can be allowed to weather naturally; other wood sidings must be protected by paint or stain.

Choosing the Siding

There are various types of siding patterns available. The interlocking styles will be somewhat more expensive than those that simply overlap, but they go up faster and form a tighter seal.

Often, lumberyards or building supply centers don't carry all types of horizontal wood siding, so you may have to look around to find what you want. You should also shop around to compare prices, since they can vary widely. If you are going to cover an entire house, ask the manager for a price reduction. It is often given for a large quantity.

Preparing the Wall

On new construction, wrap the house with building paper (15-pound felt) stapled in horizontal rows. Start from the bottom and work up, lapping each strip over the one below by 2 inches, and overlapping ends by 4 inches. Carry the felt around door and window openings.

If you are applying new siding over old, you will not need building paper. However, if the house is not insulated and has no vapor barrier, drill ½-inch holes through the siding and sheathing at the top and bottom of each stud cavity. This permits vapor to escape. Nail on 1-inch-thick sheets of rigid insulation; leave ⅛-inch gaps at the corners and top edge for moisture to escape. When re-siding, it is important to locate all the studs. You may be able to spot them if the old siding has exposed nails. If not, pry off the top piece of horizontal siding, then use your chalkline as a plumb bob and snap a vertical line over the center of each stud.

If doors and windows are wood-framed, they need flashing (see page 68).

Sight along the house to spot any bulging boards. Nail them back into place.

Applying board siding

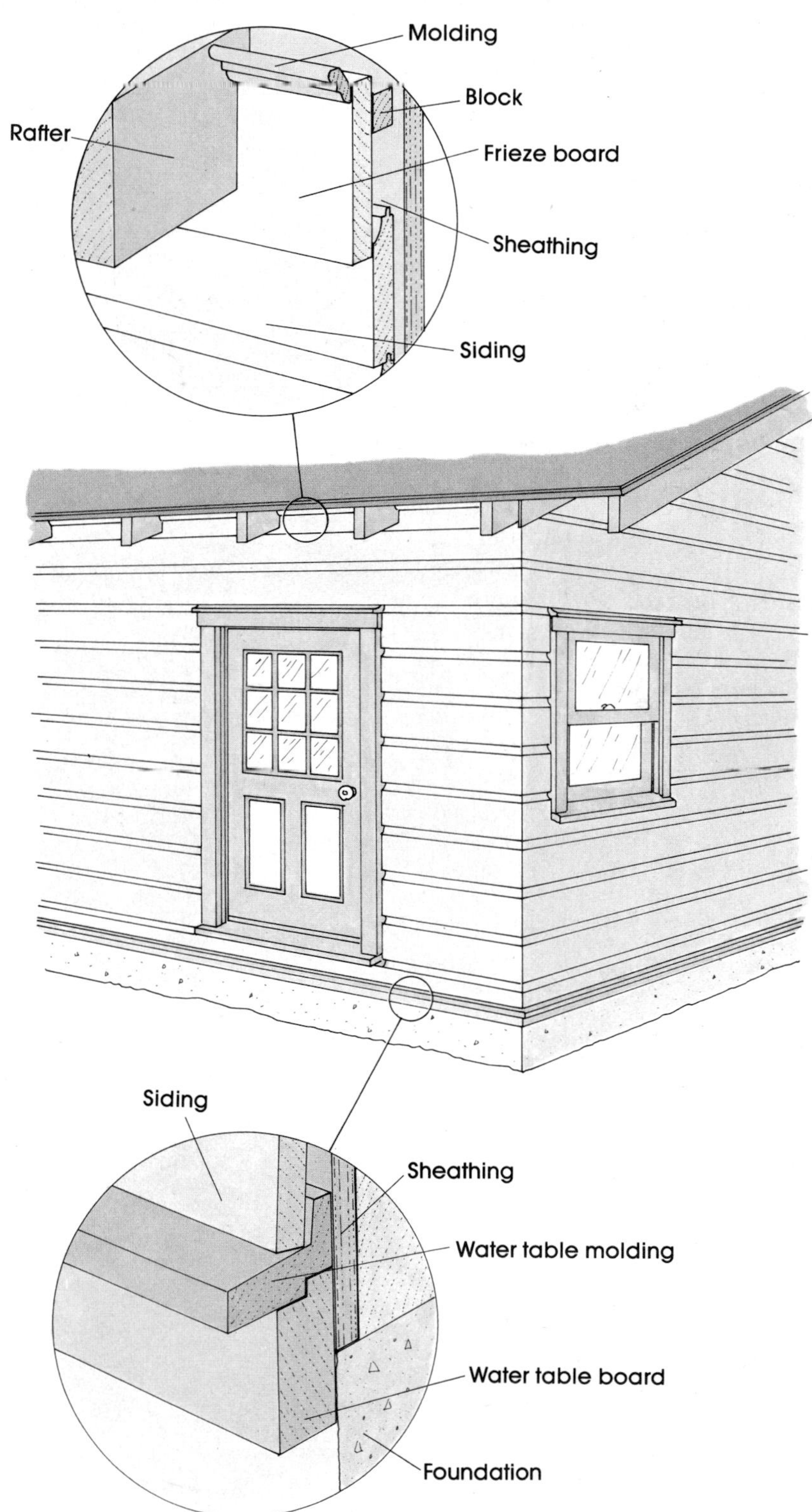

HORIZONTAL WOOD SIDING

Furring strips. If you are applying horizontal siding over an irregular surface, such as shingles, weathered vertical board siding, or concrete block, you must apply furring strips first to provide a flat nailing surface.

Place 2-by-3 furring strips around all door and window openings, and center them over all studs. If you are putting furring over concrete block, use case-hardened nails.

If the existing siding has trim board down the corners, place furring strips next to them. Trim boards for the new siding should then be nailed directly over the existing ones. For a closed corner (see box on corner treatments, opposite), place furring strips down each corner.

Here's an important tip: after the furring strips are in place, sight down the wall. If you see any inward bowing of the strips where they conform to the inward bowing of the wall, pry them out and slip shingle shims behind these areas until the surface is straight. Unevenness is easy enough to correct at this stage, and very difficult later.

Installing the Siding

Put the story pole (see page 67) in place at each corner and transfer the marks to the corner boards. Periodically check your work against the story pole marks to ensure the siding does not drift up or down.

To determine if the bottom of the existing siding is horizontal, place a level on the under edge of the bottom board. If it is horizontal, use it as a guide in placing the first board. If it isn't, use a line level on a stretched chalkline and snap a line around the house where the top of the first board will be placed.

For clapboard or hardboard siding, a starter board is needed around the bottom of the house in order to cant the first piece out like successive rows. It should be about 1½ inches wide and the same thickness as the top of the siding board. Starter board is not needed for beveled or tongue-and-groove siding.

Once the preparations are complete, align the first board and begin nailing on the siding. Siding with formed edges simply fits together. Use the story pole markings

Board siding installation

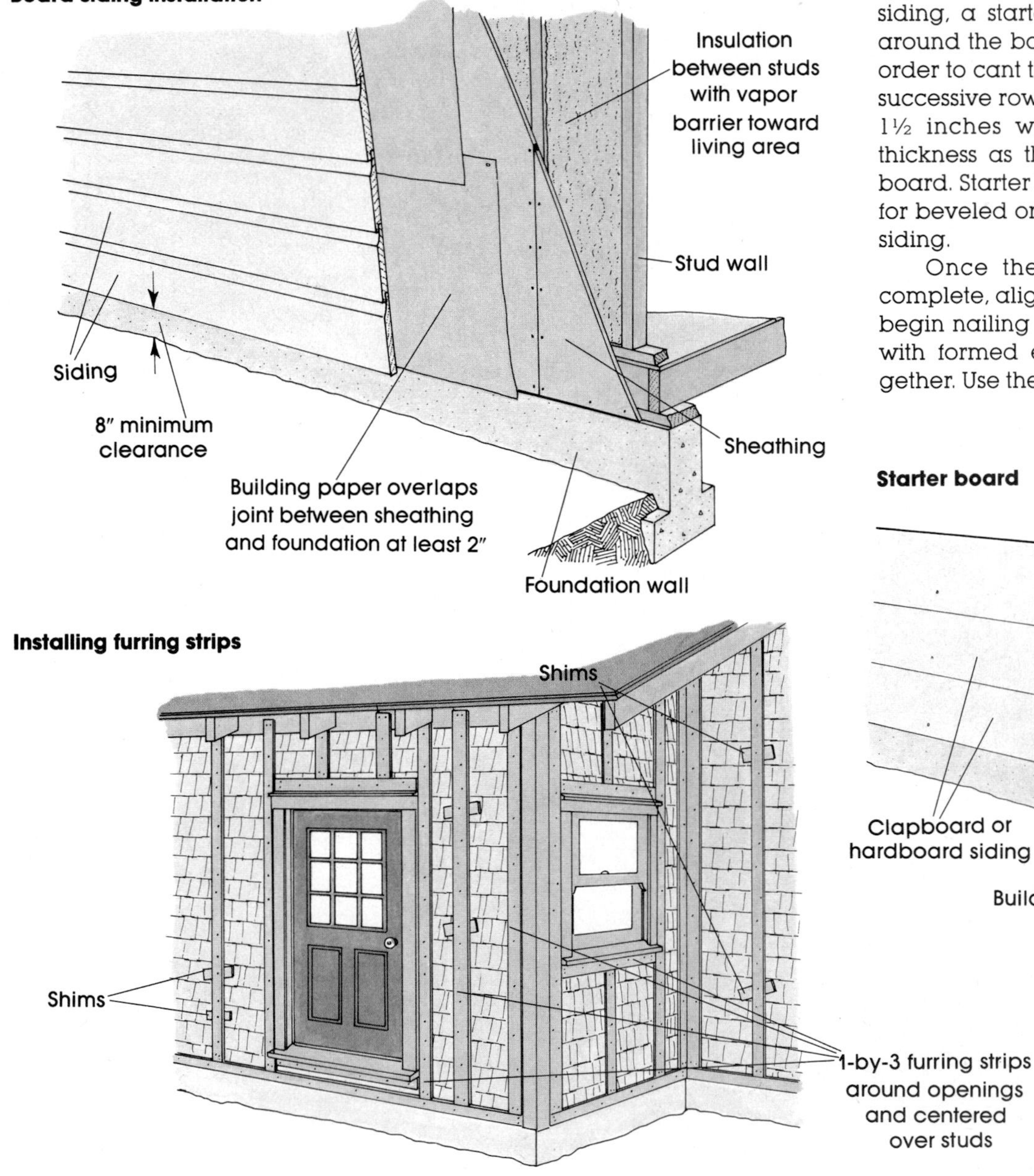

Installing furring strips

Starter board

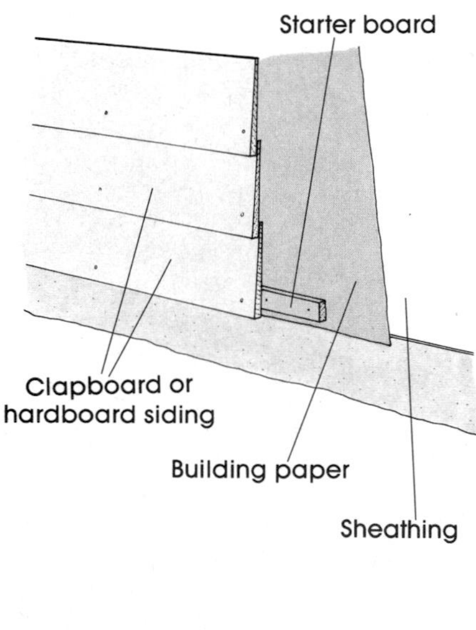

to double-check that the courses are straight. For clapboard or hardboard siding, use the story pole to keep each board straight and the exposure even.

When nailing up cedar or redwood siding that will be allowed to weather, do not use steel nails or even galvanized nails. They will rust and stain the siding with long unsightly streaks.

Instead, use aluminum nails that penetrate 1½ inches into the studs. Blind-nail tongue-and-groove up to 6 inches wide, using finish nails. For beveled and shiplap siding, and tongue-and-groove wider than 6 inches, face-nail with siding nails.

Another useful tip for nailing the ends of siding boards, where splits are likely to occur: carry a small push drill with a bit one-half the diameter of the nail, and predrill the nail holes.

Gable ends. To cut the boards at an angle matching the roof slope at the gable ends, place a sliding bevel square along the rafter as shown, then transfer that angle to the board ends.

Instead of carrying the horizontal pattern all the way up the gable, many people do the gable ends in a contrasting pattern, such as vertical board-and-batten.

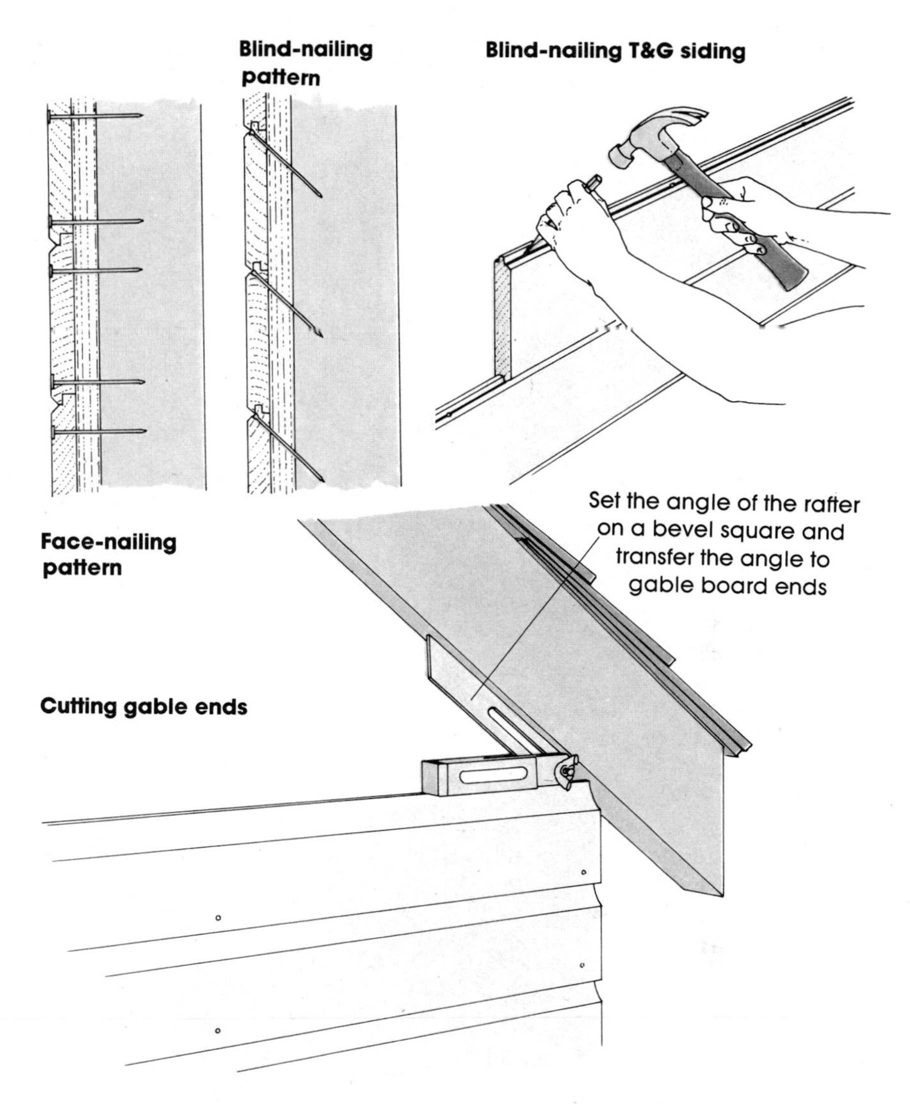

Corner Treatments

Horizontal siding requires special corner treatment to cover the exposed board ends.

For outside corners, one style uses vertical 1 by 4s overlapping each other for a tight seal. Cut the siding so the ends fit smoothly and tightly against the corner boards. Caulk this joint after all boards are in place.

Another style uses metal corner pieces. Nail siding in place with the ends just meeting at the corners, then slip the metal corner pieces over them, as shown.

On inside corners, first nail a 1 by 1 or a 2 by 2 in the corner and then butt the ends of the siding boards against it. Caulk the edges after the siding is in place.

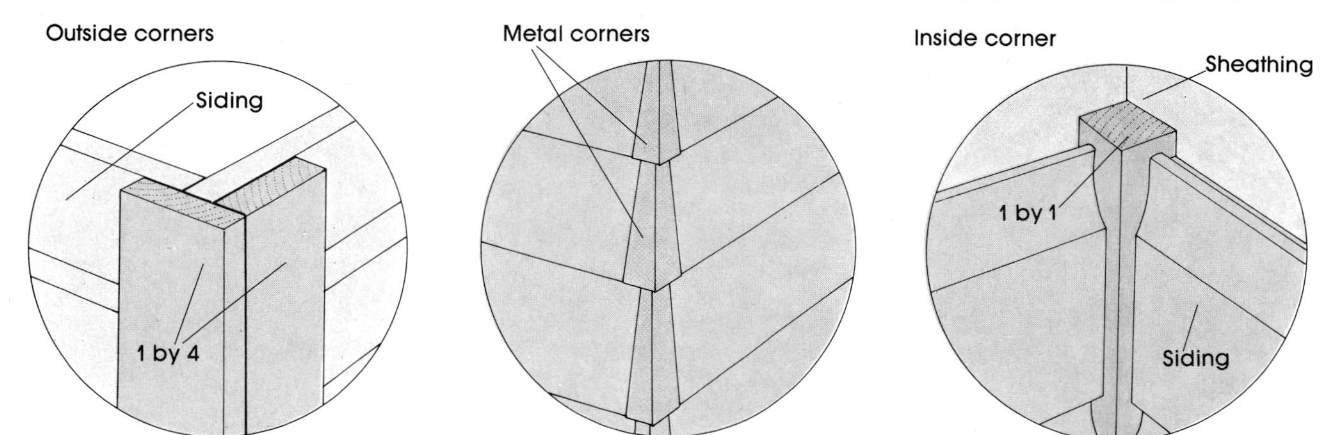

VERTICAL WOOD SIDING

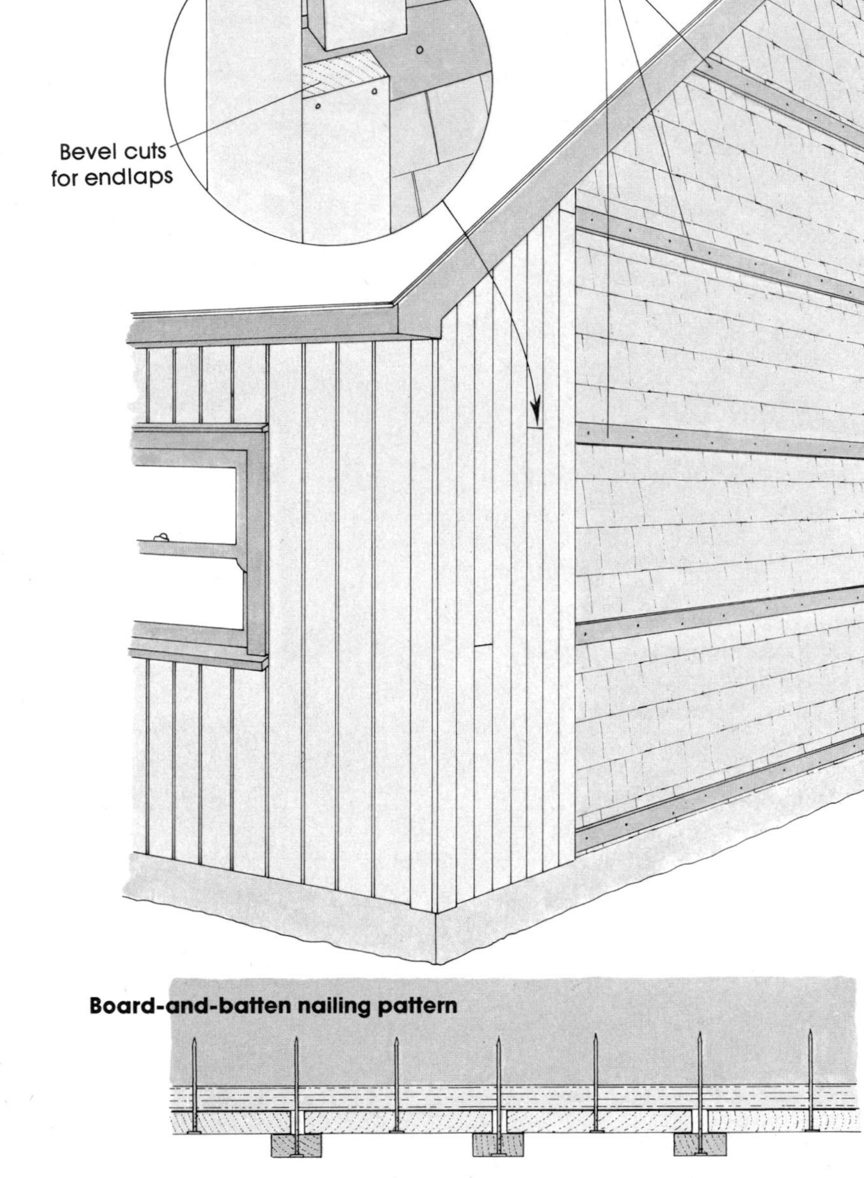

Choosing the Siding

Redwood and cedar are commonly used for vertical siding and left to weather naturally. Pine or fir can also be used if painted or stained.

The boards are usually ¾ inch thick and range in widths from 3½ inches to 11¼ inches. Some styles, such as channel siding, have rabbeted edges for a weathertight fit. Shiplap and tongue-and-groove boards are also used for vertical siding. In another common style, standard boards are nailed up vertically and the joint between each one is covered with a narrow board called a batten. Battens are generally made from 1 by 2, 1 by 3, or 1 by 4 stock; the wider battens are used with the wider boards.

Preparing the Wall

When you are applying vertical board siding over exposed studs, as in new construction or when you have torn the old siding off, blocks should be placed between the studs at 24-inch intervals to provide a nailing surface. To keep the blocks straight, snap a chalkline across the edges of the studs. The blocks will also help prevent the studs from warping, which is a common problem with "green" lumber—wood that hasn't been dried.

In new construction, or where the studs are exposed, the walls must be covered with 15-pound felt. Apply the felt from the bottom up; lap the upper strip 2 inches over the lower one, and lap the ends 4 inches. Cut and staple the felt so it covers the boards making up the rough door and window openings. Felt is not needed when putting new siding over existing siding.

If doors and windows are wood-framed, they need flashing (see page 68).

If the house has no insulation, this is a good opportunity to add it. First, to make sure that moisture from inside the house will not be trapped between the interior and exterior walls, where it might condense, drill a 1-inch hole at the top and bottom of each stud cavity. Then nail on the rigid insulation, leaving a ⅛-inch gap at each corner as an additional method of allowing any moisture to escape.

Furring strips. When covering walls that have an irregular surface, such as uneven concrete block walls or just old walls that are no longer straight, use furring strips to provide a smooth nailing surface for the new siding. Place the 2 by 3 strips around all door and window openings and in horizontal lines about 24 inches apart. Before you start to apply the siding, sight down the strips for any inward bowing. Use shingle shims behind the furring strips in these areas to straighten the wall.

Installing the Siding

When you are ready to nail up the siding, start at one corner and use a level to check that the first strip or board is perfectly vertical when nailed. Keep it vertical even if the building is out of plumb, because the problem can be hidden with trim when you finish the siding.

In applying a tongue-and-groove siding, place it with the grooved edge along the corner of the building. Tap succeeding boards into place, fitting a piece of scrap over the edge on which you are tapping to protect it. Blind-nail boards up to 6 inches wide; face-nail those wider (see page 71).

For board-and-batten, leave a ¼-inch gap between the boards—they will swell when damp and may buckle. Follow the nailing pattern illustrated at left to nail on the boards first, and then add the battens.

Where vertical boards must be end-lapped, bevel-cut the ends, as shown, to prevent water infiltration.

Covering gable ends. Find the slope of your roof by placing a sliding bevel square along the side of the house and adjusting the bevel to the angle of the rafter. Transfer this to siding that must be cut to fit the gable ends.

Gable ends with vertical siding can be done in several different ways. You can carry the siding all the way to the roof line in a continuous sweep, as is commonly done with narrow redwood or cedar strips. Or, depending on the material you are using and the style you want to set, you could put horizontal siding over the gable ends or shingle it.

Corner treatments. The standard corner treatment for vertical siding is to use overlapping 1 by 4s, or a 1 by 4 overlapping a 1 by 3. Caulk should be applied along the building corners before the corner pieces are nailed on.

Gable treatments

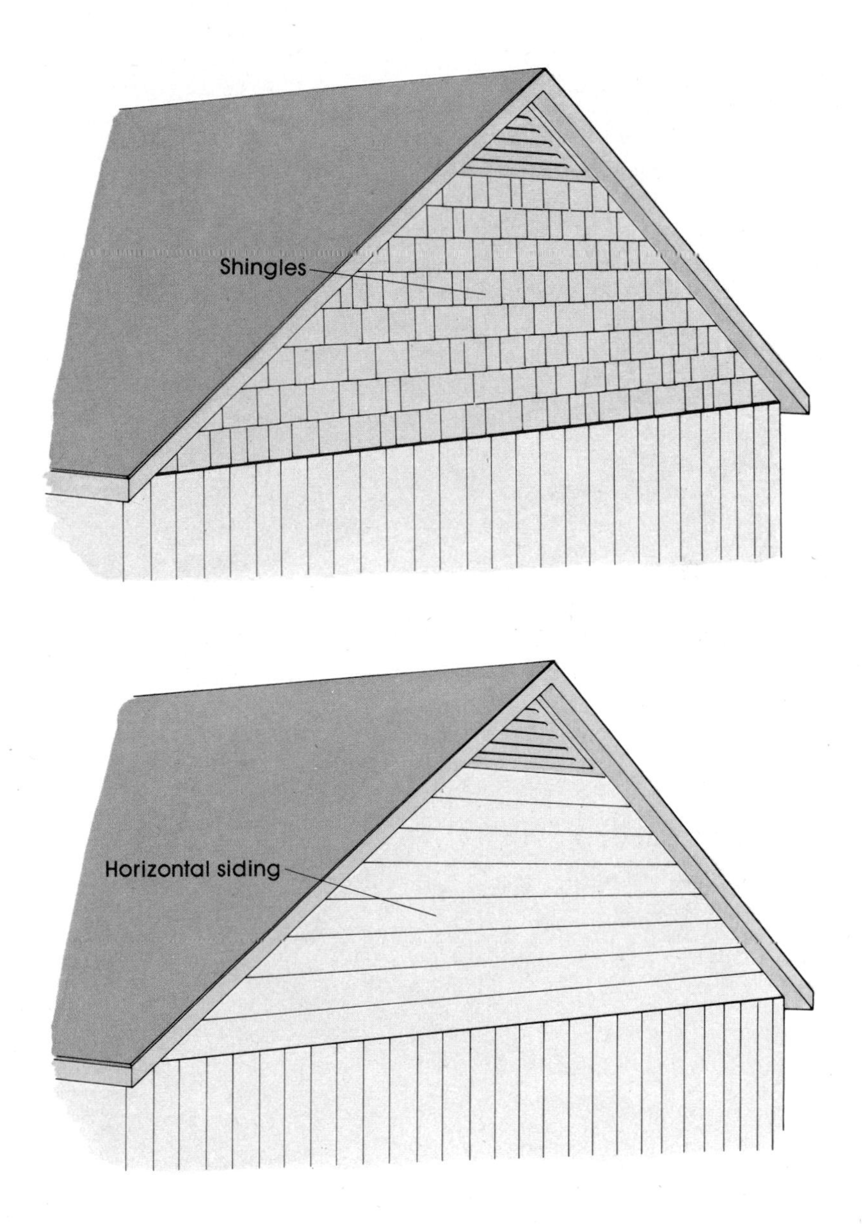

Trimming corners

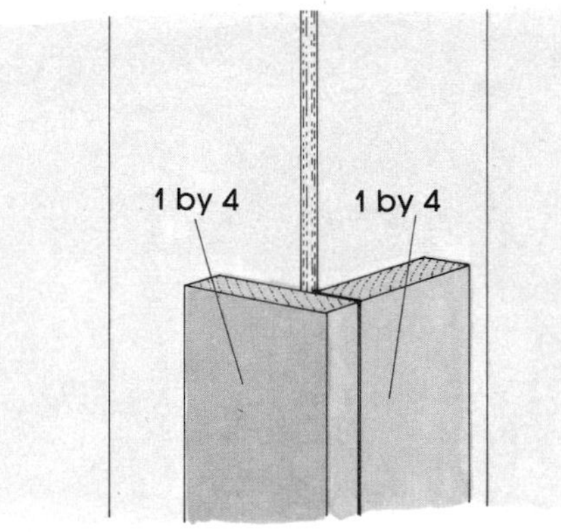

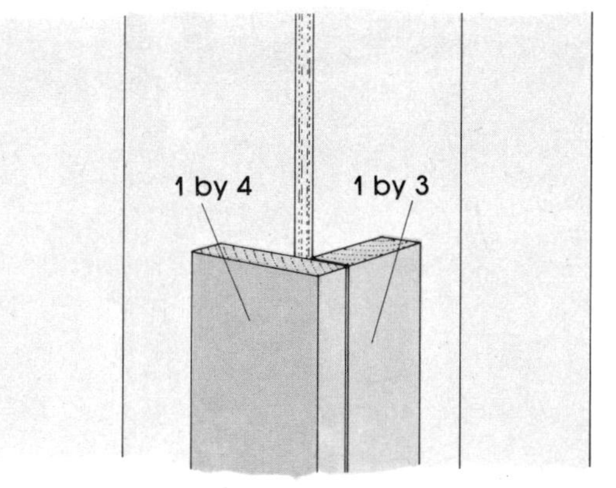

PANEL SIDING

Choosing the Panels

Plywood and hardboard panels are widely used for exterior wall covering because they are relatively low-priced and go up quickly. The panels come in a wide variety of patterns and designs. Standard sizes are 4 by 8, 4 by 9, and 4 by 10. The bottom edge of the panel should overlap the foundation wall by at least 2 inches, but should not be closer than 8 inches to the ground. For houses framed with standard 8-foot-high walls, this means 4-by-9 panels are needed.

For new installation, panel siding should be not less than ⅜ inch thick; patterned or grooved material usually must be thicker. When re-siding over irregularly surfaced siding such as shingles or horizontal siding, the thicker and stiffer ½-inch or ⅝-inch siding should be used; over smoother surfaces, such as old panels or board-and-batten (with battens removed), you could use ⅜-inch panels.

Installing the Panels

In new installation, for panels ½ inch or less thick, use 6d galvanized nails. For thicker siding, use 8d galvanized nails. In re-siding, use galvanized nails that will penetrate 1½ inches into the studs. Nails must be placed 6 inches apart around the edges and 12 inches in the field, or interior, portion.

Leave a ⅛-inch gap between panels to allow for expansion. When cutting around doors and windows, leave a ¼-inch gap to make the fitting job easier. Gaps around doors and windows should be caulked before trim is applied.

In new construction, there is no need for felt under the siding. But flashing paper is required around window and door openings, and some builders put a strip of flashing paper down studs where panels meet. If doors and windows are wood-framed, they need flashing (see page 68).

Plywood and hardboard panels must be cut with the exposed face down because circular saws leave a jagged edge on the upper surface. So all measurements marked on the panels must be transferred in mirror image to their backs. This can be a little confusing at first; double-check your markings and think before you cut.

Panel siding is heavy and awkward to handle, so have a helper on hand. After the panel is positioned, one person can hold it while the other nails it in place.

Panels are normally placed vertically. The leading edge of each panel must fall in the center of a stud. Proper positioning of the first panel is critical, since all successive panels must fit smoothly and vertically against it. If the first one is slightly out of line, succeeding panels will become increasingly out of line. If you have difficulty getting the first one square, spread your corrections over several panels.

For panel styles that overlap, position the first panel so the rabbeted edge faces out. When the next panel is put in place, the inward-facing rabbet fits over it.

Here's a tip for use in new construction: apply panel siding before installing rafters, with the tops flush with the top of the cap plate, and then cut the rafters to fit over the siding. This way you avoid having to notch the panels to fit around each rafter.

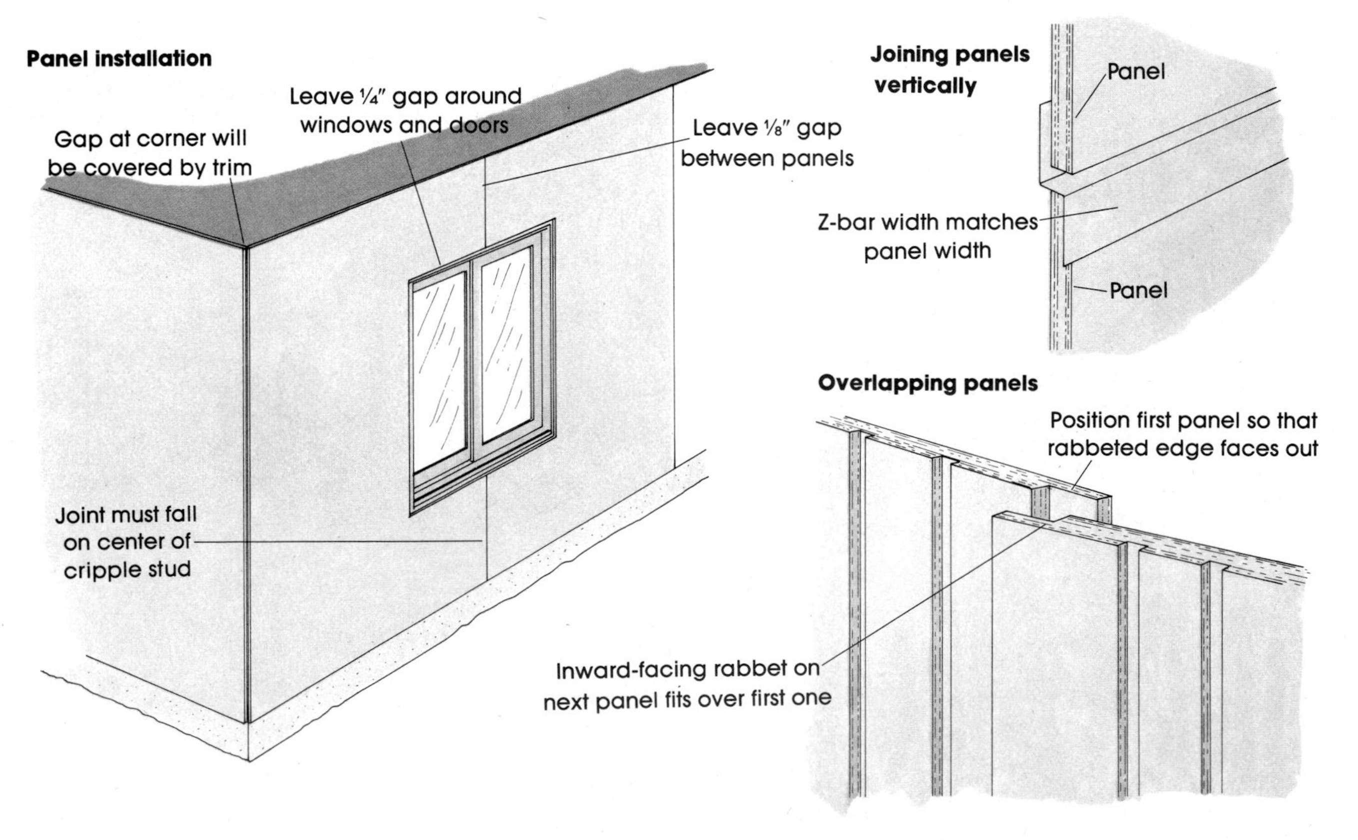

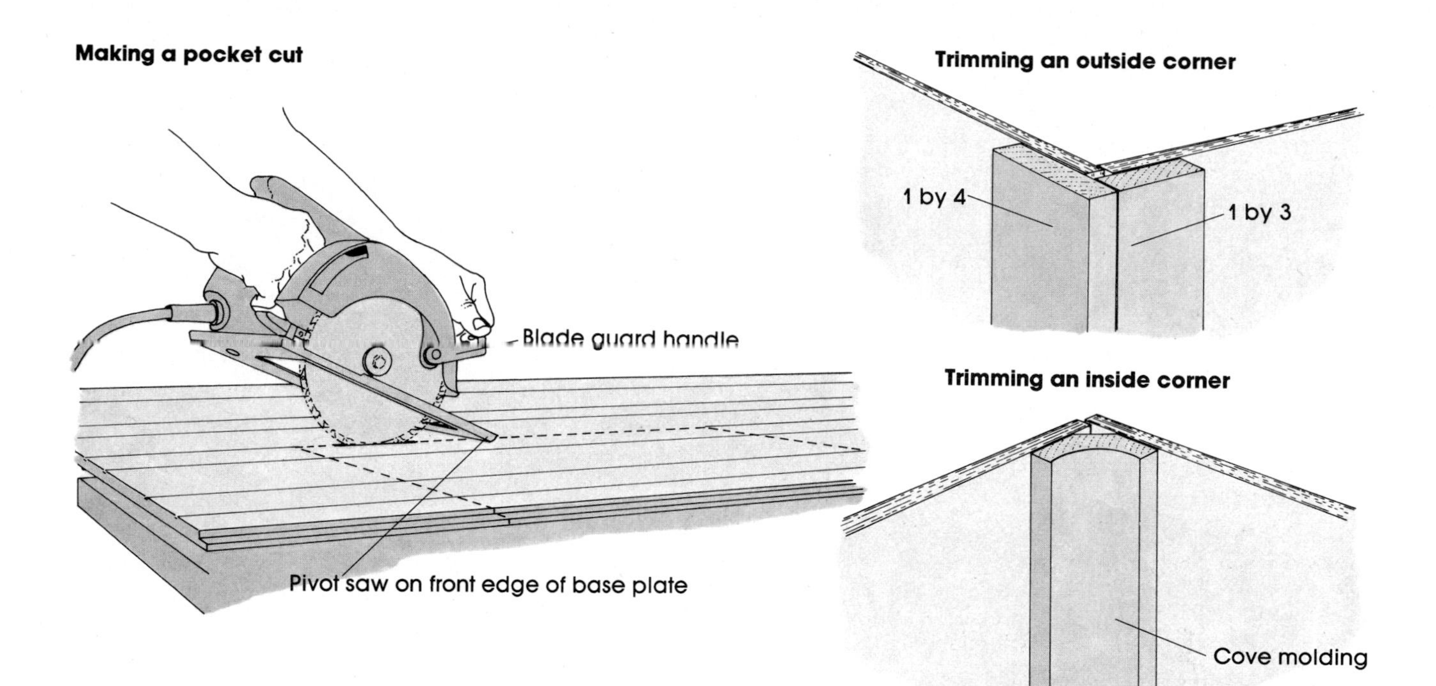

Cutting around openings. When working around openings, use full sheets of plywood, rather than trying to make a patchwork of small leftover pieces. No matter how far the panel extends around the opening, the leading edge must always fall in the center of a cripple stud (a partial stud found above the opening and, for a window, also below.)

Careful measuring is required here to avoid expensive mistakes. Always leave a ¼-inch gap to make it easier to fit the panel around an opening. Measure from the leading edge of the last panel to ¼ inch from the edge of the opening. Measure from the top of the panel down to the top of the opening, again leaving ¼-inch clearance. Measure from the top of the opening to the bottom plus ¼ inch. If the panel fits completely over the opening, measure from one side of the opening to the other side, plus ¼ inch.

Lay out these measurements on the panel with a straightedge, then cut. To make an inside cut, or when cutting an opening in the middle of a panel, use a "pocket cut." Do this by positioning the circular saw at one end of the line with the blade over the line. Raise the blade guard on the saw and start the saw. Gently lower the saw into the wood, then cut along the line. With practice, the starting cut will just intersect the right-angle cut line. If you miss, do not back the saw up, as it will likely jam and possibly jerk backward. Instead, finish the cut with a handsaw.

Trim work. When the panels are in place, caulk the gap between the panels and the window or door jamb, and between panels meeting at the corners.

Outside corners are normally trimmed with two 1 by 4s that overlap as shown. Some people like to use a 1 by 3 on one side and a 1 by 4 for the overlapping piece of trim for more symmetry. Inside corners can be done in the same manner, or you can nail in a length of cove molding, as shown.

Windows and doors are commonly trimmed with 1 by 4s, as illustrated. Note that the top piece usually overlaps the side trim rather than meeting it at a 45-degree angle, so as to minimize the chance of water running underneath. Caulk the edges of the trim. Flashing above the trim is not generally used with plywood siding.

Cover the tops of the panels with lengths of 1 by 4 where they meet the rafters or frieze blocks.

If siding must be fitted over a water pipe and the hose bibcock (spigot) cannot be removed, mark the pipe position on the panel and then drill a hole there ¼ inch larger than the pipe diameter. From the bottom of the siding, cut out a strip the same width as the pipe diameter. Slide the siding in place over the pipe, then glue the strip back into the slot. Caulk the opening around the pipe.

Gable ends. When paneling the gable ends of a house, first cover the top edge of the plywood across the end of the house with Z-bar flashing. Order the flashing according to the thickness of your siding. The bottom edge laps over the top of the siding and the top edge fits behind the gable end siding.

To calculate the gable end cuts, measure the distance from the top of the plywood at the corner to the bottom of the rafter, less ¼ inch clearance. Measure 4 feet along the end of the house, then measure up from there to the bottom of the rafter, less ¼ inch. Pencil these measurements on a sheet of plywood, then connect the tops of the long and short lines, which will be the line of the roof.

After the gable end siding is up, caulk the gap between the tops of the panels and the rafter and cover it with trim boards.

SHINGLE SIDING

Choosing the Shingles

Apart from sheer attractiveness, cedar shingles have several advantages over other siding: they are long-lasting and require no painting, and you can do the job alone. Although shingles are somewhat more expensive than other types of siding, you will save on maintenance. Two types of cedar shingles are commonly used: red cedar, which weathers to a silvery gray, medium brown, or dark brown, depending on local climatic conditions, and white cedar, which weathers to a silvery gray. They are sold in grades 1, 2, and 3.

In addition to the standard shingles sold, there are fancy-cut shingles, which allow you to create a variety of patterns in the shingling. They are generally grade 1, and expensive. They are applied in the same manner as standard shingles.

Shingles are sold in lengths of 16 inches, 18 inches, and 24 inches. The maximum exposure should be ½ inch less than half the overall length, which works out to 7½ inches, 8½ inches, and 11½ inches. These exposures can be reduced for better protection and to make courses line up with the story pole layout (see page 67).

Calculating Your Needs

Wood shingles are sold in bundles made up according to the length and number of shingles. To order enough for your house, all you need to know is the square footage of your walls. Find this by multiplying height times length of each wall, then subtract the square feet of each window from this total. From this figure, your supplier will quickly estimate your needs. You should allow 10 percent to 15 percent extra for waste. You can return any unused bundles, but you might keep one on hand for later repairs.

Preparing the Wall

Shingling on walls is normally applied over solid sheathing. The corners and all door and window openings should be covered with kraft paper to protect against possible water infiltration. However, over the wall sheathing, use what is called red resin paper. It is similar to 15-pound felt but is not asphalt-impregnated. It allows the shingles to breathe while still blocking wind infiltration.

Wall preparation

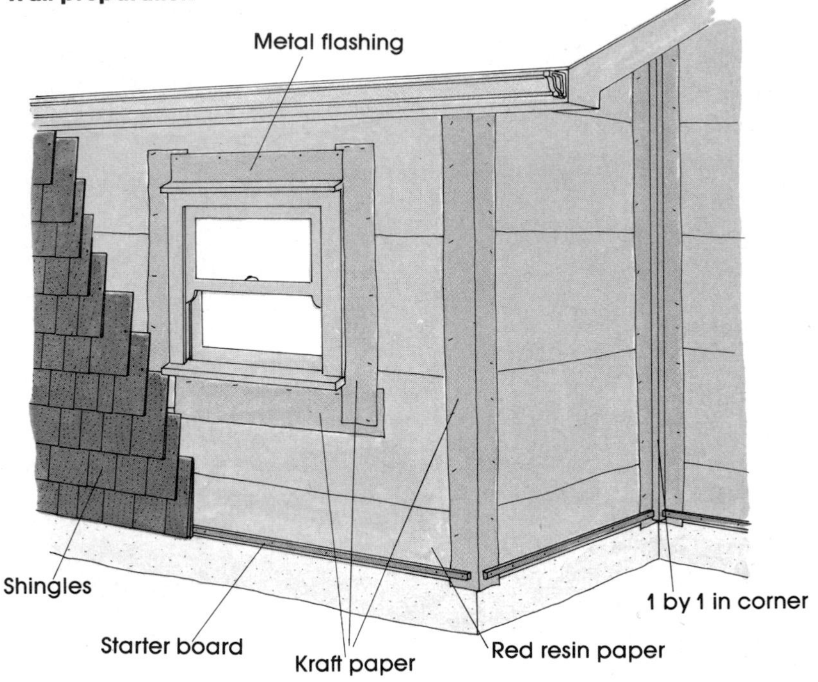

Cut and fit metal flashing over all door and window casings before the shingles are applied (see page 68). The flashing should extend 4 inches to 6 inches up the wall.

It is also advisable to paint all window jambs, sashes, and casings before the shingles go on.

Installing the Shingles

After the wall is prepared, use the story pole to lay out shingle exposures. The idea is to have shingle butts in line with the bottom of the window sill and the top of the drip cap, if possible, to minimize cutting shingles to fit.

The first shingle course across the bottom is doubled. To keep this course level, put a shingle at each corner of the building with the butt 1 inch below the sheathing. Tack a small nail to the bottom of each shingle and stretch a string between the two. Align all intervening shingles on the string line, being careful not to depress it.

For all successive courses, tack a straight 1 by 4 across the shingles in line with the story pole marks and align the shingle butts on it.

If you are shingling an older house that is not level, the first course should follow the slant of the house rather than being level; otherwise, it will emphasize the irregularity. Adjust each successive course by ⅛ inch until the courses are level. This slight change will not be noticeable.

If there is a place where you can't put the guide board between windows, snap chalklines between the story pole marks.

Fasten shingles with ring shank galvanized nails to keep them from working loose. This is particularly important when nailing shingles to ⅜-inch plywood siding. Each shingle is nailed up with only two nails, regardless of its width. Place the nails 1 inch above the butt line for the next course and ¾ inch in from the edges.

Shingles can be spaced about ⅛ inch to allow for expansion. However, many shingles are being sold "green," or freshly cut, and will shrink as they dry, so check with your dealer. No gap between shingles should be closer than 1½ inches to a gap in the course below, and no gap should be in line with one less than three courses below. When putting a course above a door or window, don't let a gap line up with the window or door edge.

Where shingles must be cut to fit around obstructions, measure and cut with a handsaw or, for curves, a coping saw. For fine trimming when fitting along casing or trim boards, use a block plane on the shingle edge.

If shingles must be shortened to fit above a window or door opening, trim the shingles from the butt end. Trimming along the top will mean thicker shingles under the row above, causing a bulge.

Corner treatments. Where shingles meet at corners, they can be mitered, woven, or butted against trim boards.

Mitering is the most painstaking method. Each shingle must be fitted against another, and the edges miter-cut with a power saw. Although they are very good-looking, mitered corners are the least effective at blocking wind-driven rain.

Woven corners provide better weather protection and are more commonly used, but they too must be individually fitted and then cut.

To weave corners, nail the bottom layer of the doubled starter course around the bottom of the house. Now start the next layer of shingles. Consult the illustration to see how the top shingle on side A extends beyond the corner. Put the side B shingle against the extended shingle and trace its outline along the back of the side A shingle. Cut along that line with a keyhole saw and nail the shingles in place.

On the next row up, repeat the process, this time extending the end shingle on side B beyond the corner and, on its back, marking the outline of the shingle on side A. Continue up the wall in this manner, with courses overlapping in alternate rows.

The most effective weather protection—and the fastest way to shingle corners—is to use corner trim boards. These should be at least 1 by 3 or 1 by 4 stock, but for a more pronounced effect use

Wood shingle application

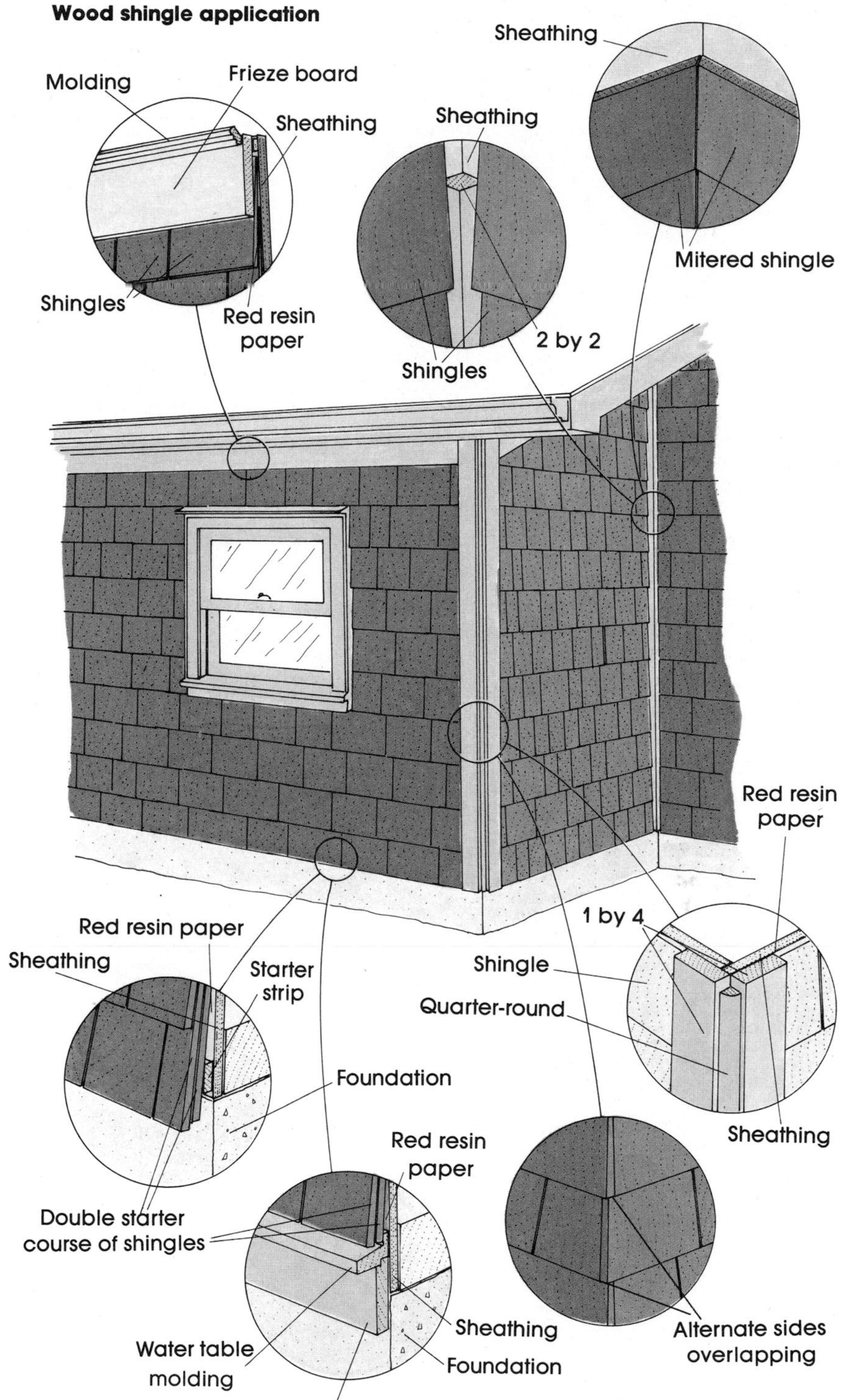

2 by 3s or 2 by 4s. You can use redwood or cedar to weather naturally with the shingles, or paint the corner boards for contrast.

Place the boards on the corners with the edges overlapping to form a tight seal, then bring each course of shingles flush with the board edges. Use a plane to trim shingle edges where necessary for a smooth fit against the board.

Inside corners can also be woven, or the shingle edges can be butted against a 2 by 2 set.

Put a bead of caulk between the shingles and corner boards.

APPLYING STUCCO SIDING

Stucco is one of the more widely used sidings in this country because of its durable, weather-tight shield and low maintenance. But it has some drawbacks. It is hard to apply, and it may crack if poorly applied or if the house settles. Nonetheless, stucco can be applied by the careful do-it-yourselfer. However, consider finding a pro to work with you or at least get you started.

Stucco is a mix of cement, sand, lime (to keep it plastic), and water. You can buy premixed sacks of the dry ingredients or mix your own. Stucco is applied in three separate coats over metal lath or chicken wire nailed to the sheathing. Special nails with fiber washers, called furring nails, are used to space the wire away from the building paper. The first coat, called the scratch coat, is about ¼ inch thick. It is supported on the wall by the metal lath or wire. Before the first coat dries, it is scored with a rake-like trowel called a scarifier so the second coat will adhere to it. The second coat, called the brown coat because years ago it contained brown sand, is then troweled on about ⅜ inch to ½ inch thick. This coat must be put on smoothly and evenly because the final finish coat, which is no more than ¼ inch thick, will not hide any mistakes.

Weather is an important consideration in applying stucco. If a wall has long exposure to hot afternoon sun, the stucco may dry too rapidly, then shrink and crack. This is particularly true with the thick second coat. If you have such a wall, plan your stuccoing for times of the day when the direct sun will be minimal, or hang shade cloths from the eaves.

The best temperature range for applying stucco is between 50 degrees F and 80 degrees F; it should not be applied when temperatures drop below 40 degrees F, because it becomes too stiff to work properly.

Tools for Stuccoing

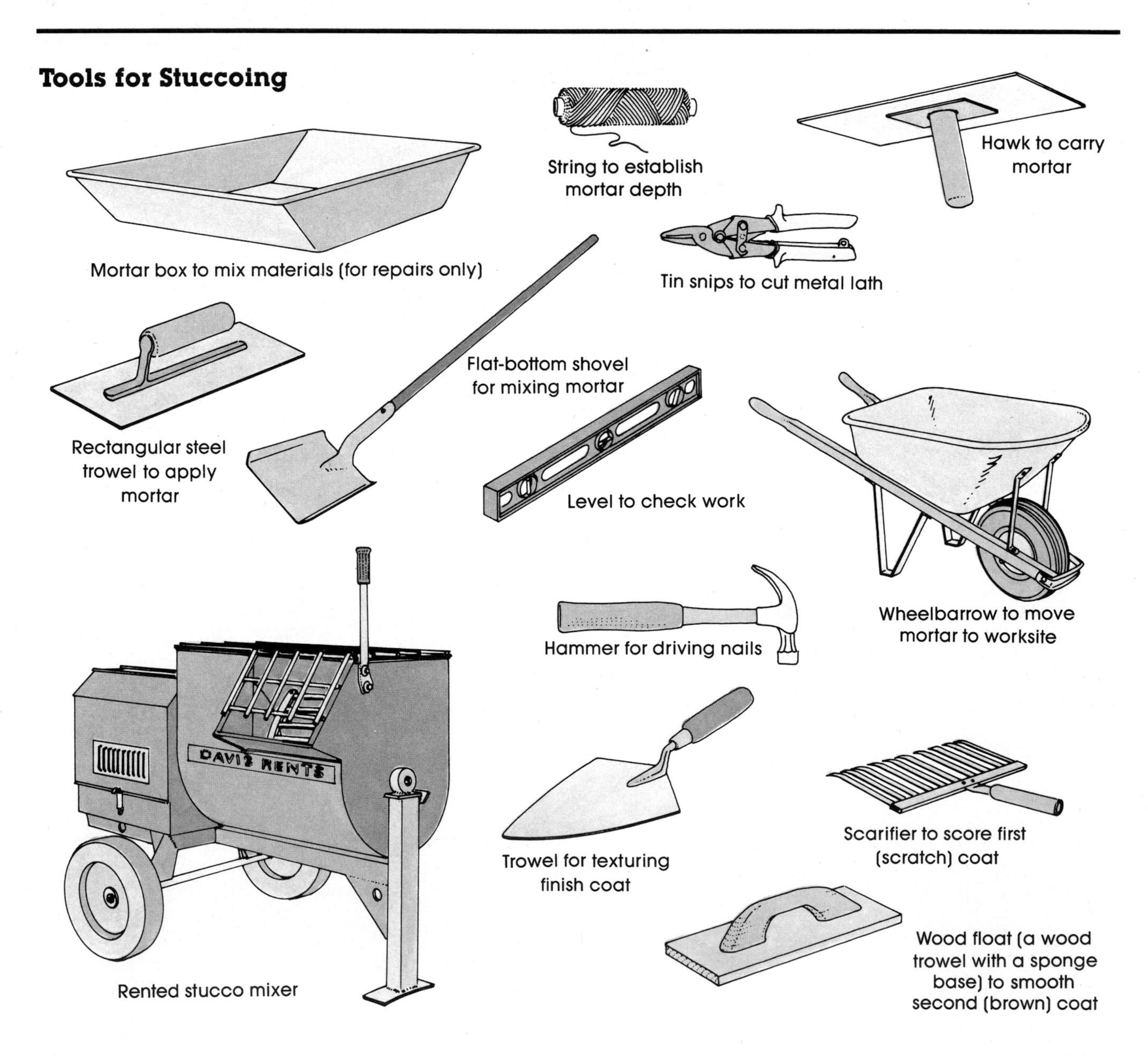

Stucco application

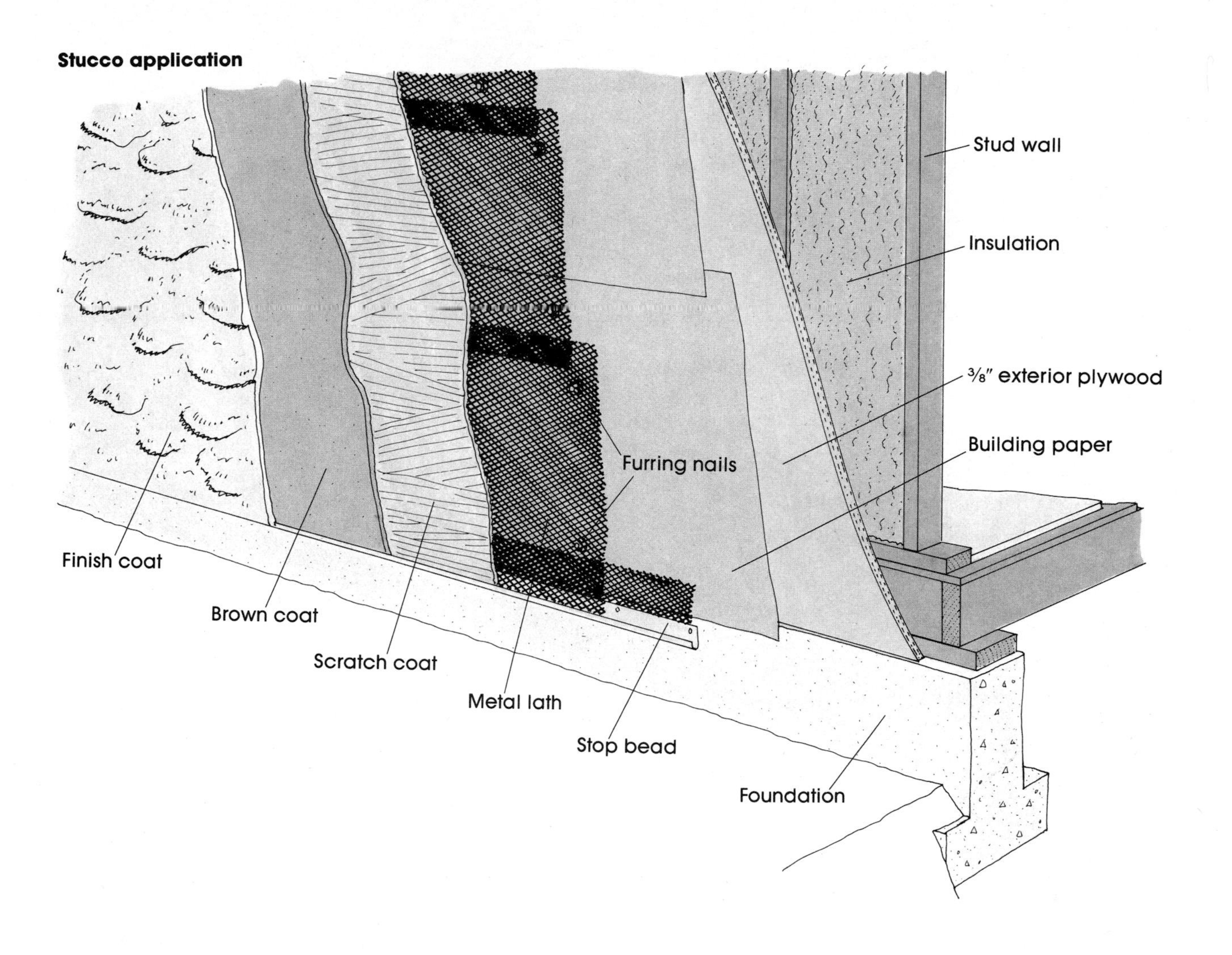

Preparing the Wall

In new construction. Cover the walls of the house with the cheapest exterior-grade 3/8-inch plywood available. Cover this with building paper (similar to 15-pound felt) in horizontal strips overlapping 2 inches top to bottom and 4 inches end to end.

Now snap a level chalkline around the foundation about 4 inches from the ground. Along this line nail the reinforcing wire called stop bead, mesh side up. Use case-hardened masonry nails spaced every 8 inches.

Nail the stop bead around window and door casings every 6 inches cutting it to length with tin snips. Keep the mesh pointing out. Now nail the stop bead around the top edge of the wall with the mesh pointing down.

Applying stop bead

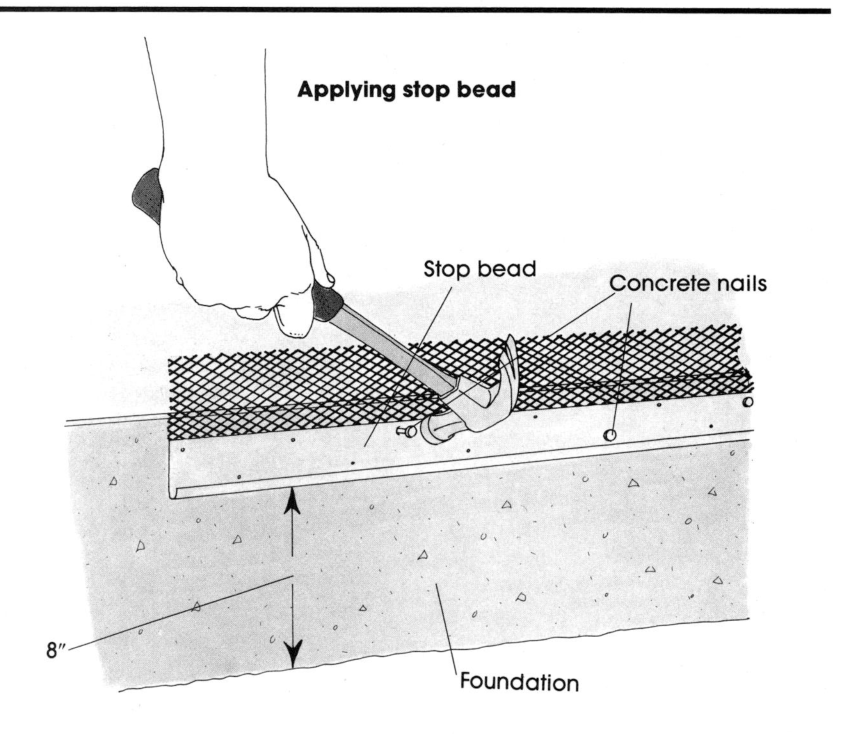

APPLYING STUCCO SIDING
CONTINUED

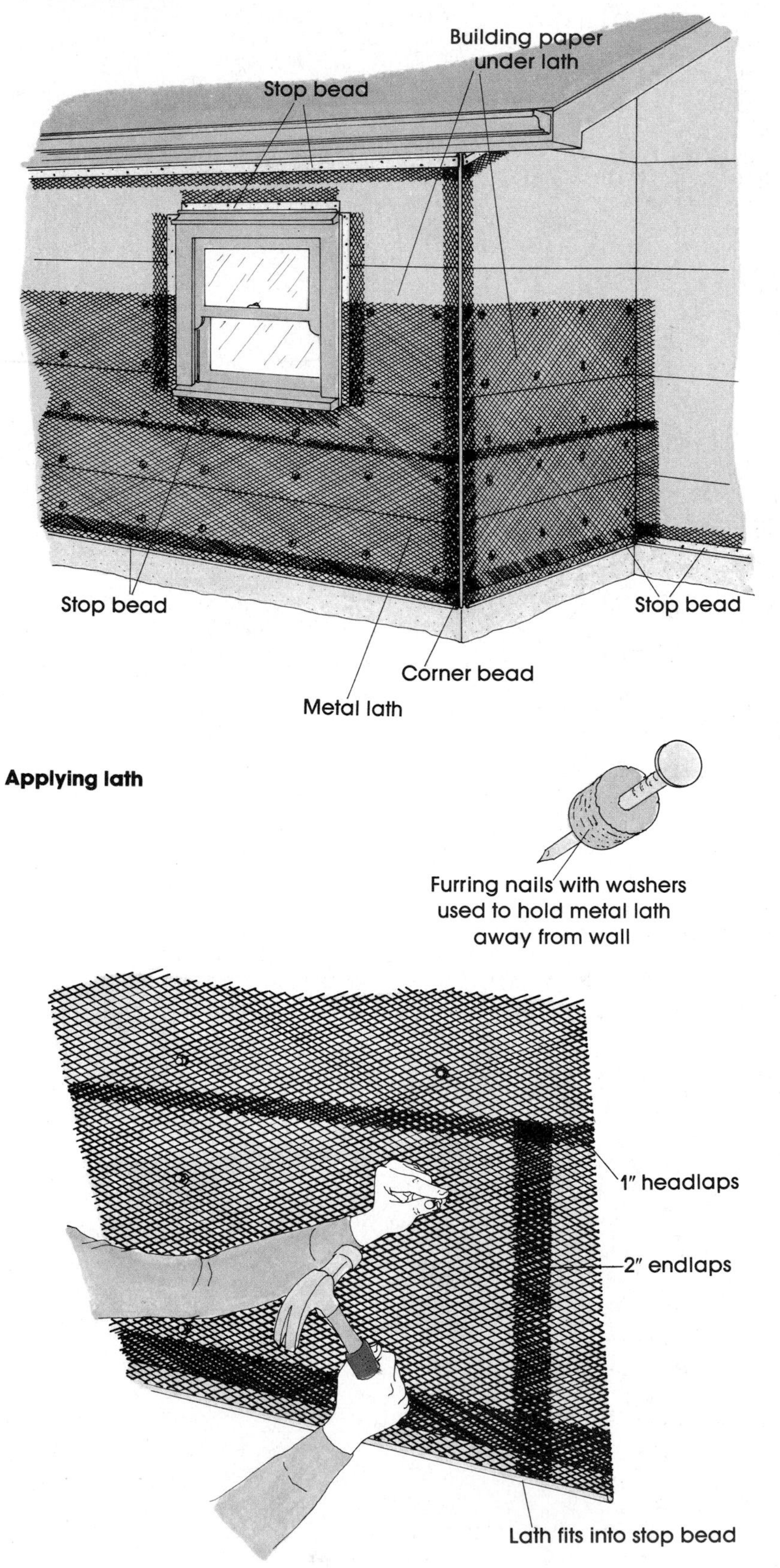

Starting from the bottom at one corner and working up, nail on the metal lath sheets. Note that the nails are sold with the washers close to the nail head. Push the washers away from the nail head, hook the wire between the nail head and washers, then drive the nail home. Stretch the wire mesh each time as you hook the nail. Note that the bottom of the first sheet rests in the stop bead.

Overlap the strips 1 inch where they meet horizontally and 2 inches at the ends. Place furring nails every 6 inches on the lath. Cut the sheets to fit into the stop bead around window and door openings.

At the corners, nail a strip of corner bead vertically along the edge. The metal strip should protrude from the edge by ¾ inch. Use a level to make the bead vertical, and if more than one length must be used, carefully align the two ends. Double-check that the beads are accurately placed—they will be important guides when you apply the second coat.

With the metal lath all in place, the wall preparation is complete. Apply masking tape to door and window casings to protect them from stains.

Over masonry or old stucco. Stuccoing a masonry wall or stuccoing over old stucco does not require felt or the metal lath, nor is the first, or scratch, coat of stucco used.

First scrub the wall thoroughly with a wire brush to remove all dirt and loose or flaky material. Spray the wall with a hose. After it dries, use a roller to coat the wall with a masonry bonding agent, available at hardware or paint stores. Let it dry overnight, then go directly to the brown coat.

Mixing Stucco

Stucco must be mixed, just like the concrete that it is. The easiest way is to buy stucco mix and rent a stucco mixer big enough to handle one full sack. What you pay in rent will be more than made up in time. Dump in the mix and add water until it is a soft plastic consistency that you can squeeze and hold in your hand without any drips.

For small repair jobs, you can mix your own. You'll need a large metal mortar box, which you can either buy at your local hardware store or rent. Or you can rent a stucco mixer.

A standard mix is 3 parts building sand (as opposed to the finer mortar sand), 1 part Portland cement, and ¼ part lime. The amount of water will depend on how wet the sand is.

Put all the dry ingredients in the mortar box and turn the combination repeatedly with a flat-bottom shovel until it is an even color. Push the mix into one-half of the box and slowly add water while you keep mixing. When it looks almost right, be careful, because at this point you can make the mistake of adding too much water. The stucco should be a plastic consistency, neither dry nor soupy. If it is too soupy, add equal proportions of the dry ingredients until it is right.

One caution in new construction: lumber is often sold "green" and will shrink as it dries. This can cause the stucco to crack. To prevent this, use lumber graded "dry." Or you can apply the scratch coat to seal the house, then wait two or three months to apply the brown and finish coats.

Applying the Mortar

Once the mortar is mixed, take it to the work area in a wheelbarrow. It's now ready to be placed on a hawk and then troweled into the wall. If you are inexperienced and work slowly, keep the mortar covered with a piece of plastic to prevent its drying too fast. If it appears dry, stir it around but do not add more water.

You can place mortar directly on your hawk with a trowel, or keep a pile of it on a mortar board made from a sheet of scrap plywood placed over a couple of sawhorses near the wall.

Load the mortar board, then chop the mortar with the edge of the trowel and spread it evenly around the center of the mortarboard. To load your trowel, cut through the mortar on the side away from you, then simultaneously tilt the mortar board down toward you while scooping the section of mortar off the board with the trowel, as shown. A few practice tries, while everyone watches you drop the mortar on the ground, will help you perfect this move. To keep the pile centered, turn the hawk a quarter turn after each trowel load; continue to take the mortar from the side away from you.

Apply the mortar at the top of the wall and work downward. As you press the mortar against the wall, tilt the top of the trowel slightly away from the wall and then apply the mortar with an upward sweep, pressing it into the metal lath.

Continue working your way downward, blending adjoining areas together with smooth horizontal strokes.

Use this basic move in applying each coat.

Scratch coat. The first coat should be ¼ inch to ⅜ inch thick, with most of the mortar pressed behind the lath. Cover the lath evenly with just enough mortar for a faint impression to show through. Check carefully that no bulges or pockets occur along the wall.

Before the mortar dries, score it horizontally with the rake-like tines of the scarifier. The tines should bite deep enough to almost touch the lath without exposing it. The grooves created by the scarifier will make the next coat adhere.

When working on a long or high wall, be sure that the mortar doesn't dry before you scarify it. Allow the scratch coat to dry 4 to 6 hours before proceeding.

Brown coat. Before applying this coat—the most critical of the three because of the evenness required—two preparatory steps are necessary.

First, stretch strings horizontally across the wall. One string should be placed near the top of the wall, one near the bottom, and one (or more, if the wall is high) centered between the first two. Attach the strings to nails placed beyond the corners so that each string is held

Loading the trowel

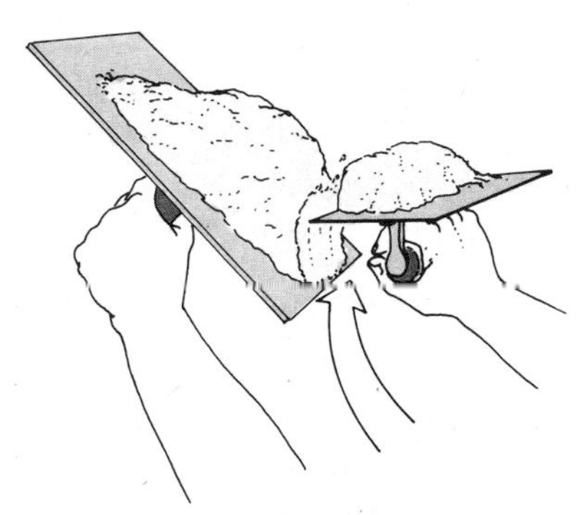

Scoop stucco onto upturned trowel

Mortar application

Press mortar in with an even upward stroke

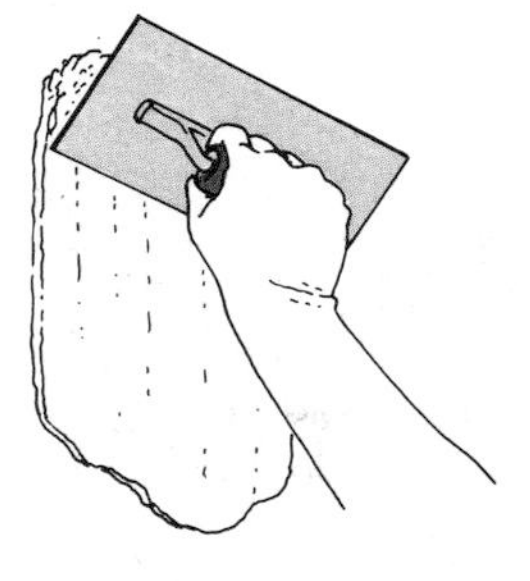

Scarifying the scratch coat

Make horizontal grooves with scarifier

APPLYING STUCCO SIDING
CONTINUED

Preparing for the brown coat

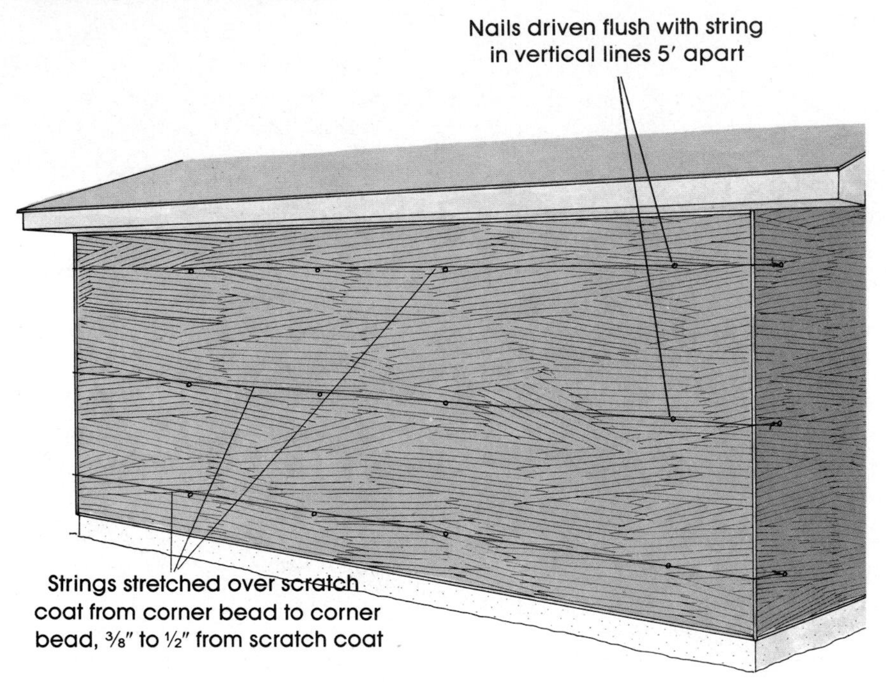

Applying the brown coat

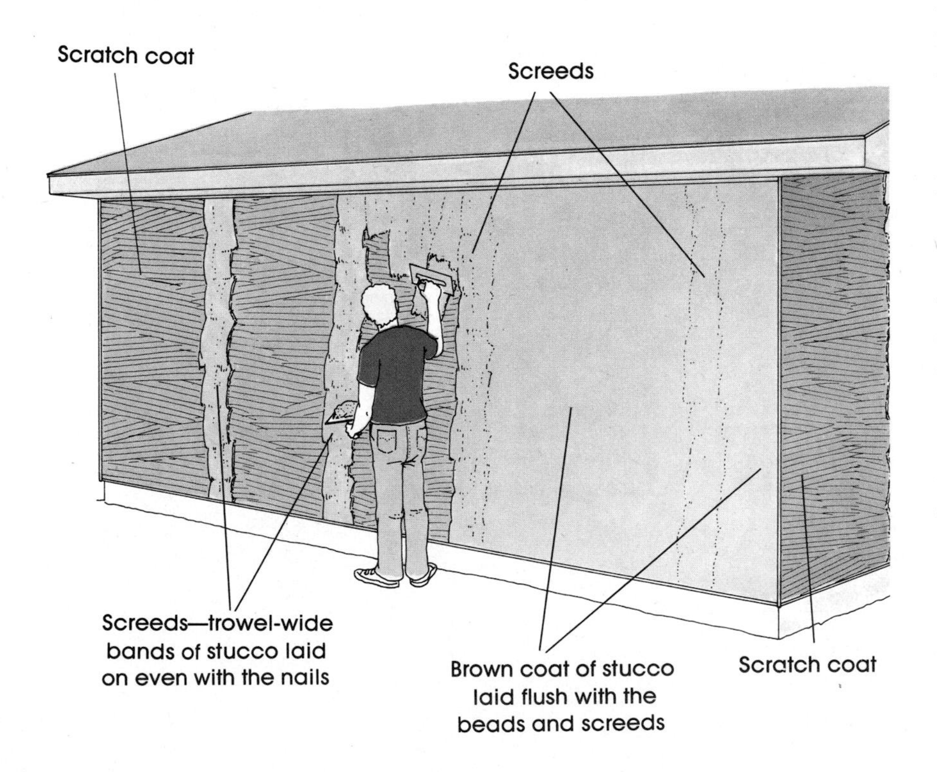

out from the wall by the protruding stop beads. This space between wall and string establishes the thickness of the brown coat. Every 5 feet along the string lines, drive a 3-inch galvanized roofing nail so the head is just flush with the inside edge of the string. The nails will form vertical lines up the wall. Place nails about 12 inches away from both sides of doors and windows to form vertical lines, spaced on the string lines. When the nails are all in, remove the strings.

The next step is to trowel on the screeds, which are narrow vertical strips of stucco applied over the nails just thick enough to cover the nail heads. When the screeds are in place and have dried 24 hours, the areas between are filled in to the same thickness as the screeds, which serve as leveling guides.

Apply the stucco screeds from top to bottom, just covering the nail heads, in strips only as wide as the steel trowel. After each strip is applied, place a straight board on edge against the screeds and smooth out any high or low spots before moving on to the next one. Let the screeds dry 24 hours.

Before applying the brown coat, spray the wall lightly with the hose nozzle set for misting. Dampen only as much wall as you can cover in an hour.

Apply the brown coat from top to bottom. Feather this coat into the screeds with smooth, even strokes so no joint is visible. Smooth the mortar over the beads at the corners, top, and bottom.

Regularly check your work—sight down the wall and place a straightedge vertically along it. Use the straightedge in a sawing motion to shave off high points; fill in low spots with more mortar.

Allow each section between a pair of screeds to dry for about an hour, then go over it with a wood float in circular motions. Rub lightly until you feel the sand in the brown coat just work to the surface. Let the brown coat dry for at least 24 hours, misting it gently every 4 hours or so to keep it from drying too rapidly. Protect it from hot direct sun with shade cloths or wet sheets draped on the wall to help hold in moisture.

Finish coat. The final coat, only 1/8 inch to 1/4 inch thick, can be colored and textured according to your tastes. For the various shades of brown, tan, or off-white, pigment available from your stucco supplier is mixed in the final coat. For a white finish, use white mortar and white sand.

The standard smooth-finish coat is troweled on just like the other coats and then, while still wet, is "floated" with a wood float. Go over it while it is still damp, working the float in a circular motion without leaving a pattern.

Many other patterns, as illustrated, can be obtained when applying the finish coat.

The widely used Modern American pattern is obtained by scraping the finish coat in vertical strokes with a block of 2 by 4 just after the surface moisture disappears. Press firmly enough to roughen the mortar without tearing it.

For the travertine pattern, jab the bristles of a whisk broom into the mortar while still damp, working horizontally across the wall. Then go back over it with the steel trowel in long horizontal strokes that smooth the rough top edges of the mortar.

The spatter-dash finish, commonly seen on houses over 30 years old, is created by dipping a large brush in mortar that is mixed thinner than normal, then snapping it against a stick to spatter the wall. Spray the wall with mortar in this fashion as evenly as possible. Let it dry for about one hour, then go back over it again until an even stippled effect is obtained.

For the rough effect of old English stucco, first lay on a smooth, thin layer of finish coat. After it has dried for about an hour, use a round-nosed trowel to apply irregular gobs of mortar, as shown.

Applying the finish coat

Standard smooth finish

Float with wood float

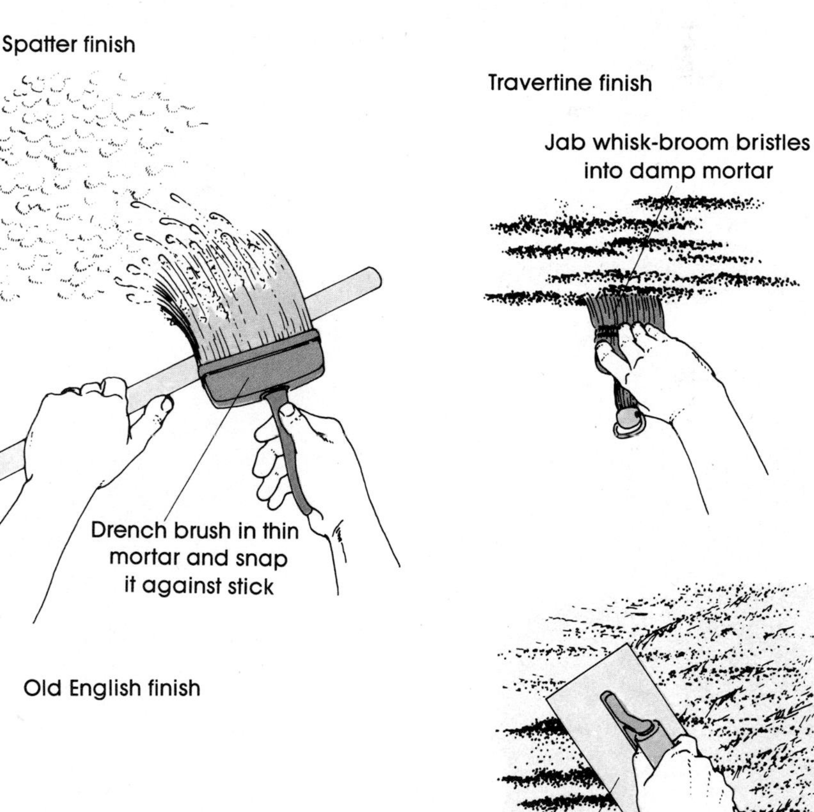

Modern American finish

Scrape surface with 2 by 4

Old English finish

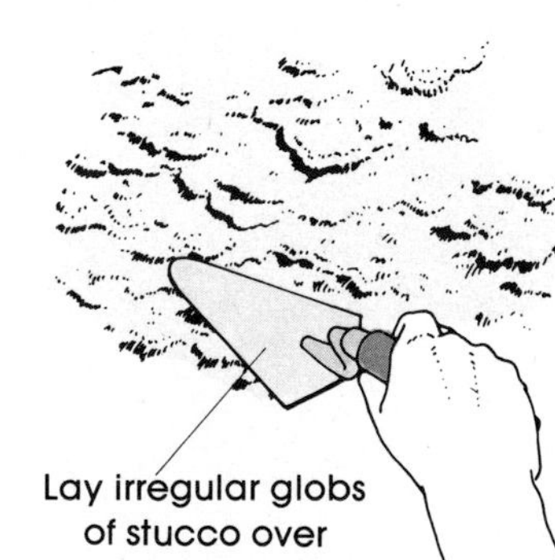

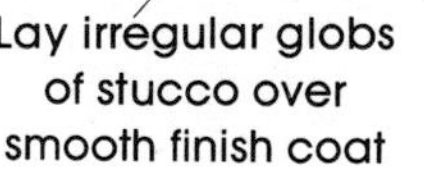
Lay irregular globs of stucco over smooth finish coat

MAINTENANCE & REPAIR

Keeping your roof and siding in good repair is a major part of maintaining the value of your home. This chapter will help you spot problems—in the roof, flashing, gutters, downspouts, or siding—and show you how to solve them before they develop into disasters.

The "skin" of your house—its roof and siding—is under constant attack by the elements. This assault persists day and night, summer and winter. The wind never quits trying to blow shingles loose; the rain hammers at the siding, washing away paint and working its way through minute cracks to dark places where rot begins; the sun's heat causes cracks to develop and materials to become dry and brittle; ice can form along the edge of the roof and cause water to back up under shingles.

One of the best defenses against this barrage is a good maintenance program. Prudent homeowners know the importance of looking after small problems before they become big problems. However, any problem, big or small, can seem insurmountable if you don't know how to solve it. For instance, a shingle blows loose on your roof. You know that unless you fix it, more shingles will follow. A professional roofer wants $30 just to fix one loose shingle. You might know it simply needs nailing down, but you worry that nailing will leave a hole in the roof, with leaks to follow. What a quandary! But wait—with hammer in one hand and this book in the other, you can fix that loose shingle yourself. Repairs are a big part of owning a house, and repairs to roofs and siding are the keys to a sound house. This chapter will help you spot problems before they develop into disasters, and show you how to solve a variety of problems.

There are a number of roof repairs that you may have to face, ranging from loose or missing roofing material to problems with valley flashing or leaks around chimney flashing.

Of course, gutters and downspouts should be cleaned every spring and fall. This is a key maintenance task, particularly if you live in a cold region, because clogged gutters can fill with water that will freeze, forming an ice dam behind which melting snow will back up under your shingles and create leaks.

Generally, siding problems are much less common, but damage and deterioration certainly can occur.

If you don't know how to make these repairs, it's difficult to get started on the job. That's what this chapter is all about—finding and fixing problems with your roof or siding before they overwhelm you.

Carefully maintained and repaired, roofs and sidings will last for many years. At left, an asphalt roof and clapboard siding are enlivened by white trim boards, fancy-cut shingles, and decorative molding.

Make an Inspection Tour

Your house is familiar to you, and because of that you may not view it with a critical eye. So go out on a walking tour around it with the idea of looking for trouble. Stand across the street and in the backyard for a good look at your roof. Use a pair of binoculars for a close look. Pay special attention to the ridge, where roofing material generally comes loose first. If you can, go up on the roof and check the flashing around the vents and the chimney. Look for cracks in the shingles around the vents, or for dried and cracked roofing cement around vent flashing.

If you have a chimney, see that the mortar holding the flashing is still sound, not loose and crumbling.

From a ladder, inspect your gutters. Look for breaks in the gutters or broken support straps. Poke the wood at the eaves under the shingles with a screwdriver to see if it's still firm. Soft wood, which almost resembles paper pulp, is a sign of rot.

Look carefully at your siding material, with an eye out for loose, split, or rotting boards or shingles. Make a list of all the potential trouble spots you've seen on your tour, then read on for the solutions.

CARING FOR ROOFS

Shingles

Asphalt Shingles

If a shingle is torn or has only minor cracks, smooth roofing cement into the damaged area. Gently raise the butt of the shingle above the damaged spot and spread some cement under there as a further precaution.

If a shingle has curled, put a coating of roofing cement under it and press it firmly back into place. Do the same for a shingle that has been blown out of position, and hold it in place with two roofing nails. Cover the nail heads with spots of roofing cement.

Missing and severely torn shingles must be replaced. A damaged shingle must be removed. In warm weather, which makes the shingles pliable, slide the straight end of a crowbar under the shingle to be removed and pry up the nails. This method has a hazard: it will lift the good shingle above the damaged one; in cool weather, the good shingle may break. A safer way is to slip a hacksaw blade under the shingle to be removed, cut each nail, and pull the shingle out.

To replace the shingle, carefully raise the one above, slide the new one in place, and nail. Put roofing cement under the edges of the new shingle and the one above to seal the repaired area.

Never attempt to replace just one tab of a torn shingle; always replace the entire shingle. However, you can make an effective (although less attractive) patch by slipping a piece of metal under the damaged shingle. Aluminum flashing, available in rolls at hardware stores, works well here. Cover the problem area with roofing cement, slip the metal in place, and nail it under the damaged shingle. Cover the nail heads with roofing cement for extra precaution. The patch will not be particularly noticeable from the ground.

Wood Shakes and Shingles

Spaces between shakes and shingles are necessary to permit them to expand when wet, but a split shingle may let water through the roof. Small cracks in wood shingles or shakes can be filled with roofing cement. For a crack more than 1/2 inch wide, slide a piece of aluminum flashing material under the shingle. Put one nail on each side of the crack to hold the flashing, then coat the nail heads and flashing with roofing cement.

A fairly common problem with wood shingles or shakes is bowing. Since rain can blow into such gaps, they should be repaired. To fix a bowed shake or shingle, split it with a wood chisel. Remove a 1/8-inch to 1/4-inch sliver along the split. Now nail the shingle down with one nail on each side of the split. Cover the split and the nail heads with roofing cement.

If a wood shingle or shake is badly cracked or rotted, it must be removed. First split it with a chisel and pull out the pieces. With a hacksaw blade, cut the nails that held the old shingle. Trim a new shingle or shake to size and tap it into place, leaving it about 3/4 inch lower than the butt line and remembering to allow 1/8-inch to 1/4-inch clearance on each side. Nail it with two nails angled so that when the shingle is driven into place, the nail heads are covered. Cover the nail heads with roofing cement.

Tar and Gravel

Flat roofs, or built-up roofs, are made of layers of heavy roofing felt and tar with a fine gravel sprinkled on the final layer of tar to keep the sun from melting it. Leaks in these roofs are sometimes diffi-

Repairing a cracked shingle

Smooth roofing cement into crack

Replacing a damaged shingle

Pry out nails

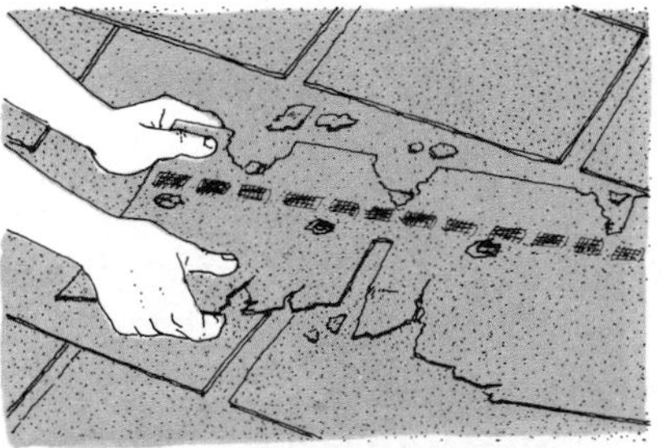

Pull out the damaged shingle

Fixing a curled shingle

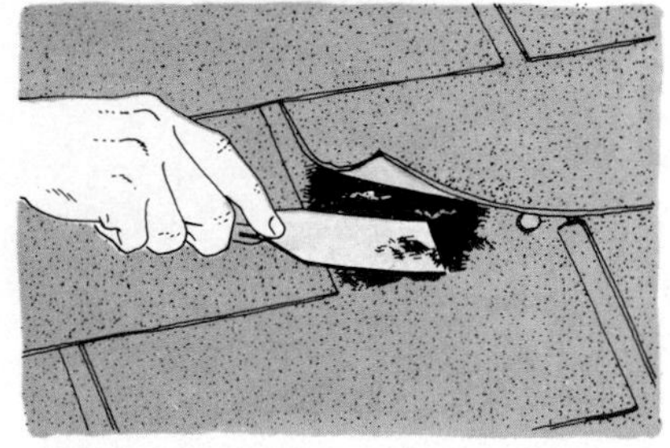

Press shingle into roofing cement

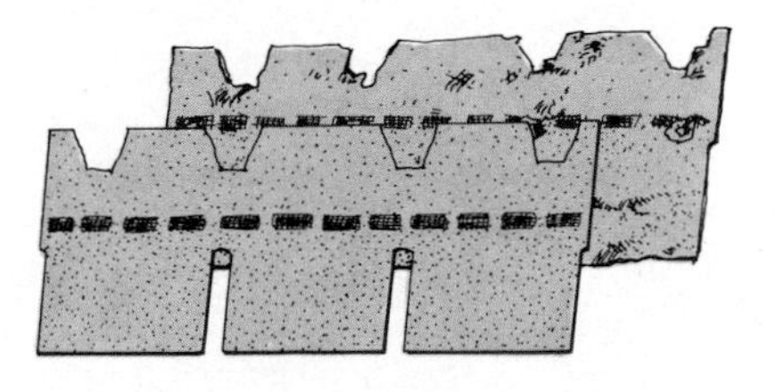

Notch the new shingle to match the tears made by nails in the old one

Nail new shingle into place

cult to spot. A bubble in the roofing material is always suspect. However, don't repair bubbles unless they are visibly broken, or unless you feel they may be the source of a leak in a low spot.

When repairing a bubble, first scrape away all the gravel around the area to be repaired. Next cut an X in the center of the bubble. Peel back each flap and liberally coat the area with roofing cement. Push the flaps back into place and cover the X and an 8-inch square around it with roofing cement. From roll-roofing or 90-pound felt, cut a square patch 6 inches larger in each direction than the X. Nail the patch down with nails spaced just 1 inch apart. Finally, cover the patch and 2 inches beyond it with roofing cement.

Two key problem areas on built-up roofs are the roof edges and the flashing. If the leak seems to be coming from either, coat it liberally with roofing cement.

If the roof still leaks, it may be too old and a new layer may be needed. Built-up roofs cannot be counted on to last much beyond ten years. If you decide to have another layer put on, get a professional crew to do it. This is dangerous work—the tar pot can explode—and should not be attempted by the amateur. Get several estimates and ask for references. However, you might consider removing the old roof yourself (see page 29) and putting on a different sort of roof. It would probably be about the same price and you would have a more durable roof.

Tile

Remove any broken tile and install a new one into its place. Refer to pages 49–51 for instructions on installing tile.

Metal and Vinyl Panels

If a panel becomes damaged, you'll need to replace the whole panel. See page 54 for instructions on installing panels.

Roll-roofing

Repair this type of roof as you would a tar and gravel one.

Repairing a bowed shake

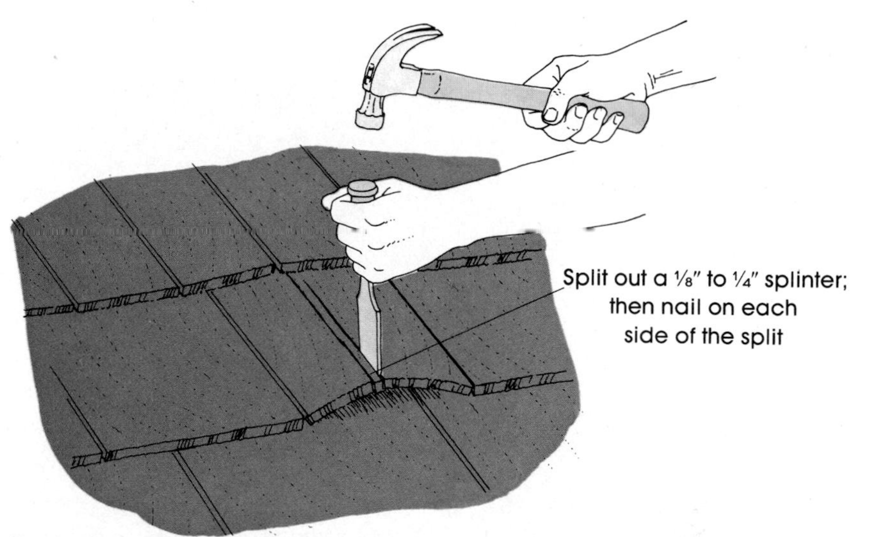

Replacing a broken shake

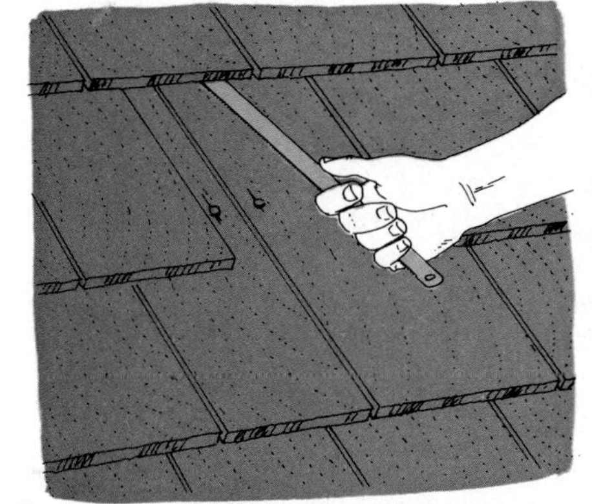

Repairing a bubble in a built-up roof

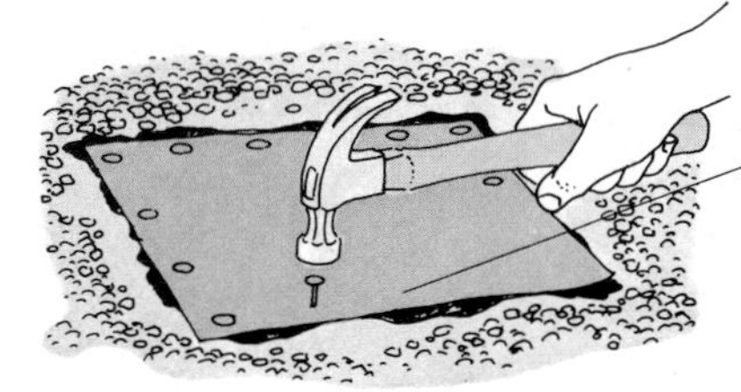

Flashing

Valleys

Leaks in valleys are sometimes the result of a break in the metal flashing itself, but that is not common. If you do see a break, coat it with roofing cement, then cover it with a piece of aluminum patching tape, available in a roll. Cut off a length to amply cover the hole, peel off the protective backing, and stick the tape in place. Its bond increases over the years.

A more common cause of leaks in valley flashing is too much water for the valley. The water overflows the edges and runs under the roofing material. To counter this, clear the valley of debris, which can act like a dam to back up water.

Next, make sure that the roofing material in the valley is cut in a smooth straight line. If part of the material, such as a shingle, is sticking out into the valley, it can act as a diverter to turn water out of the valley and under the shingles.

To really ensure against valley leaks, carefully raise asphalt shingles where they meet the valley tin and coat the area with roofing cement. Then use a cartridge gun to run a bead of roofing cement down the valley tin right next to the shingles. On the inflexible wood shingles, shakes, or tile, just run a bead of roofing cement along the edges where they extend onto the valley flashing.

Repairing cracked valley flashing

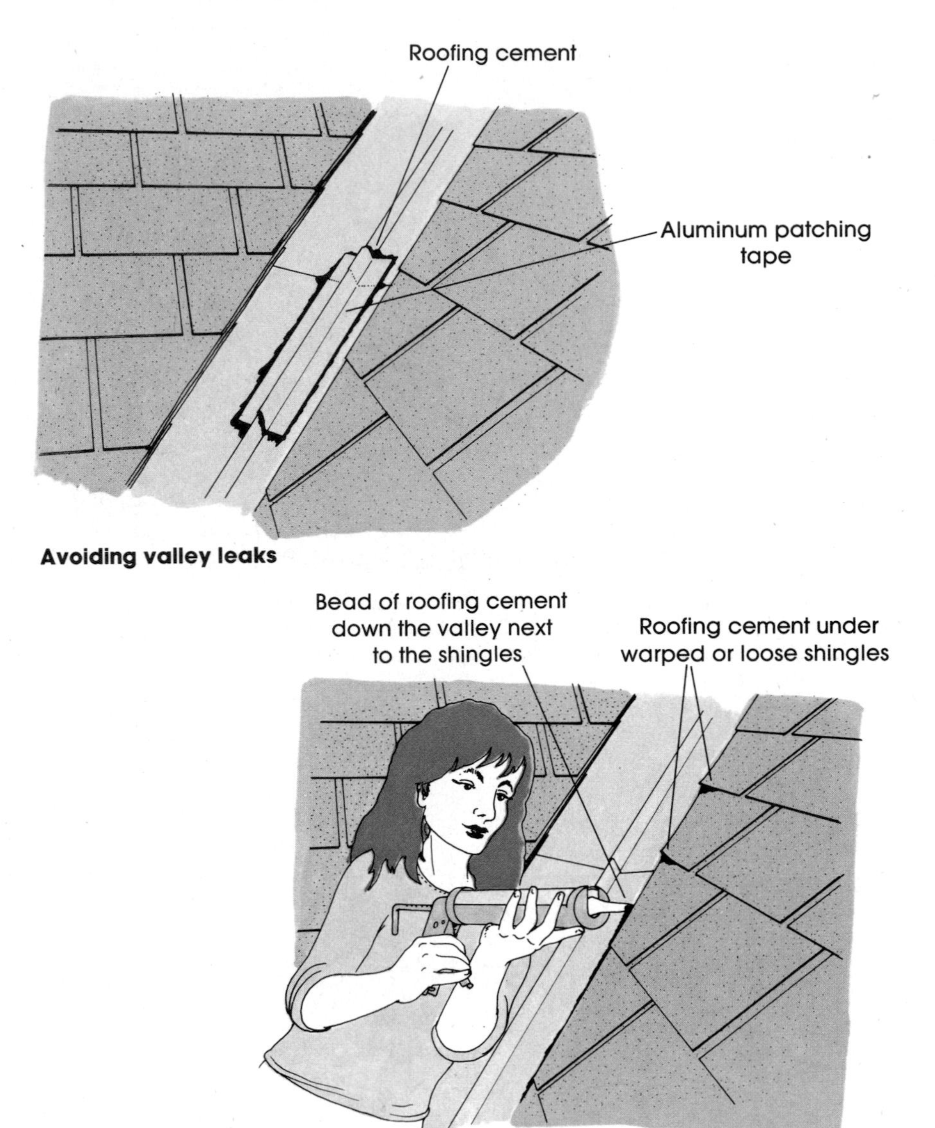

Avoiding valley leaks

Vents

Leaks often occur around vent pipes because the caulking or roofing cement has dried and shrunk or cracked. If this is the case, apply a new bead of roofing cement.

Check that the roofing surface around and above the vent pipe is in good condition. A crack in a shingle or tile will allow water to work its way under the metal flashing and down the vent pipe. It's a good practice to slightly raise the roofing material above the vent flashing and caulk underneath with roofing cement.

One final item to check is that the roofing material around the

Avoiding leaks around vent pipes

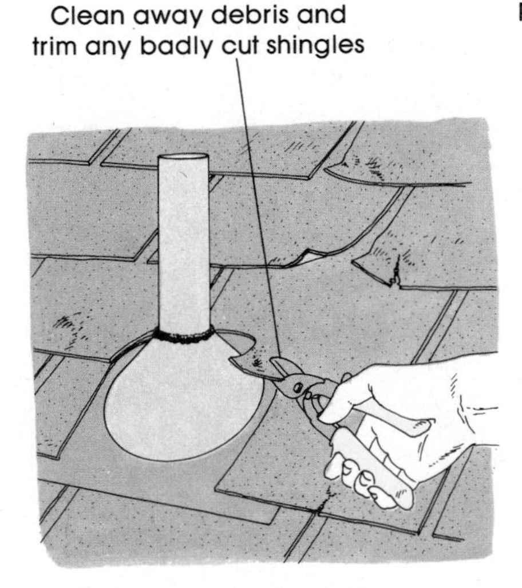

flashing is cut smoothly. If part of an asphalt shingle, wood shingle, or shake protrudes slightly, it may divert water under the shingle. If you see such poor workmanship, trim with shears (a knife might cut the flashing metal) and seal with roofing cement.

On tile roofs, check that the concrete grout (mortar) around vents has not cracked. It can be sealed temporarily with roofing cement, but it should eventually be chiseled out and replaced with new grout.

Chimneys

Flashing around the chimney is normally made of aluminum, copper, or galvanized tin. Good flashing is put down in two stages, base flashing and cap flashing (see illustration). Most frequently, leaks originate where the cap flashing is set into mortar between the bricks. Loose and crumbled mortar allows rain and snow to work behind the cap flashing. If there appears to be only a spot or two of loose mortar, scrape out the old material and fill the hole with butyl rubber caulk, which adheres well to mansonry.

If the mortar is in poor condition, break it all out with a ⅜-inch cape chisel (a narrow-tipped cold chisel) and remove the cap flashing. Remove old mortar to a depth of at least ½ inch, then clean the area thoroughly with a wire brush.

Old and new mortar do not bond well to one another. One solution is to coat the old mortar first with a bonding agent, available at most hardware stores. Another method is to soak the old mortar and the surrounding bricks with water first. This will help prevent their drawing too much moisture out of the new mortar, which would weaken it.

For the patching material, use a premixed mortar or make your own from 1 part mortar cement and 3 parts fine sand. When the flashing has been reinstalled, give yourself some added protection by coating the seams with butyl rubber caulking.

Heating Coils

If you live in an area with heavy snowfall, roof maintenance might include installing heating coils along the eaves. The coils, available at hardware stores or roofing companies, are clipped to shingles in a zigzag pattern along the eaves and are plugged into an outside outlet. The heat in the coils prevents the formation of ice dams along the eaves. Ice dams can cause water to back up under the shingles, where it can begin leaking into the house.

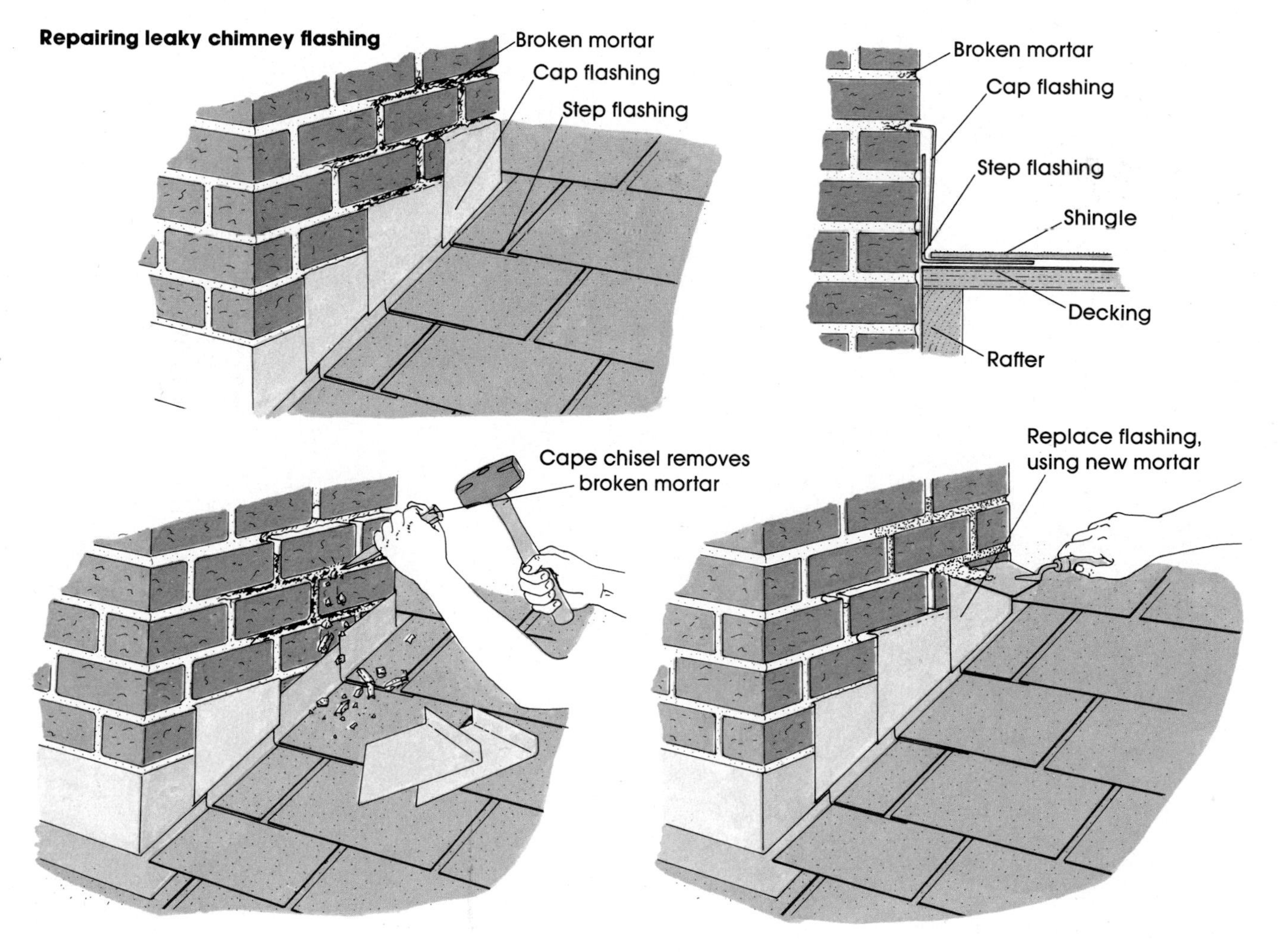

Repairing leaky chimney flashing

CARING FOR GUTTERS & DOWNSPOUTS

An important aspect of house maintenance is a good gutter and downspout system that carries water away from the house. Without such a system, water will erode the ground beneath the eaves and is likely to run into the crawl space or basement of your house, where it will encourage wood rot and termites. In addition, a gutter system in poor repair allows water and ice to back up under shingles along the eaves, which can cause leaks in the house and wood rot on the roof.

If your house does not have a gutter and downspout system, it should. New installation is covered on pages 60–61. If your house does have such a system, it should be kept in good repair.

Regular Inspections

You might be surprised how much a twice-annual inspection of your gutters will prolong their life. Each spring and each fall, clean the gutters and leaf traps of all debris. Put a hose into the downspout and see that the water runs freely.

If there appears to be a blockage in the downspout, full water pressure from the hose may clear it. If not, run a plumber's snake, the kind used for cleaning clogged sewer lines, into the downspout.

After cleaning, stand back and visually inspect the gutters and downspouts. Check that leaf traps are in place. If you live in a freezing climate, make sure that the ice did not misalign the drain system. Check that all the joints are still tight and there are no sags. Standing water in a gutter indicates sagging at that spot. Gutters should drop 1 inch for every 20 feet of run. Look carefully for loose nails or hanger straps on both the gutters and the downspouts.

Common Repairs

After cleaning the gutters, hose them out and inspect for damage. Here are some common problems and their solutions:

Sagging gutters. A sagging gutter often indicates a broken support. First, prop the gutter back into place by wedging a long 2 by 4

Two ways to clear a downspout

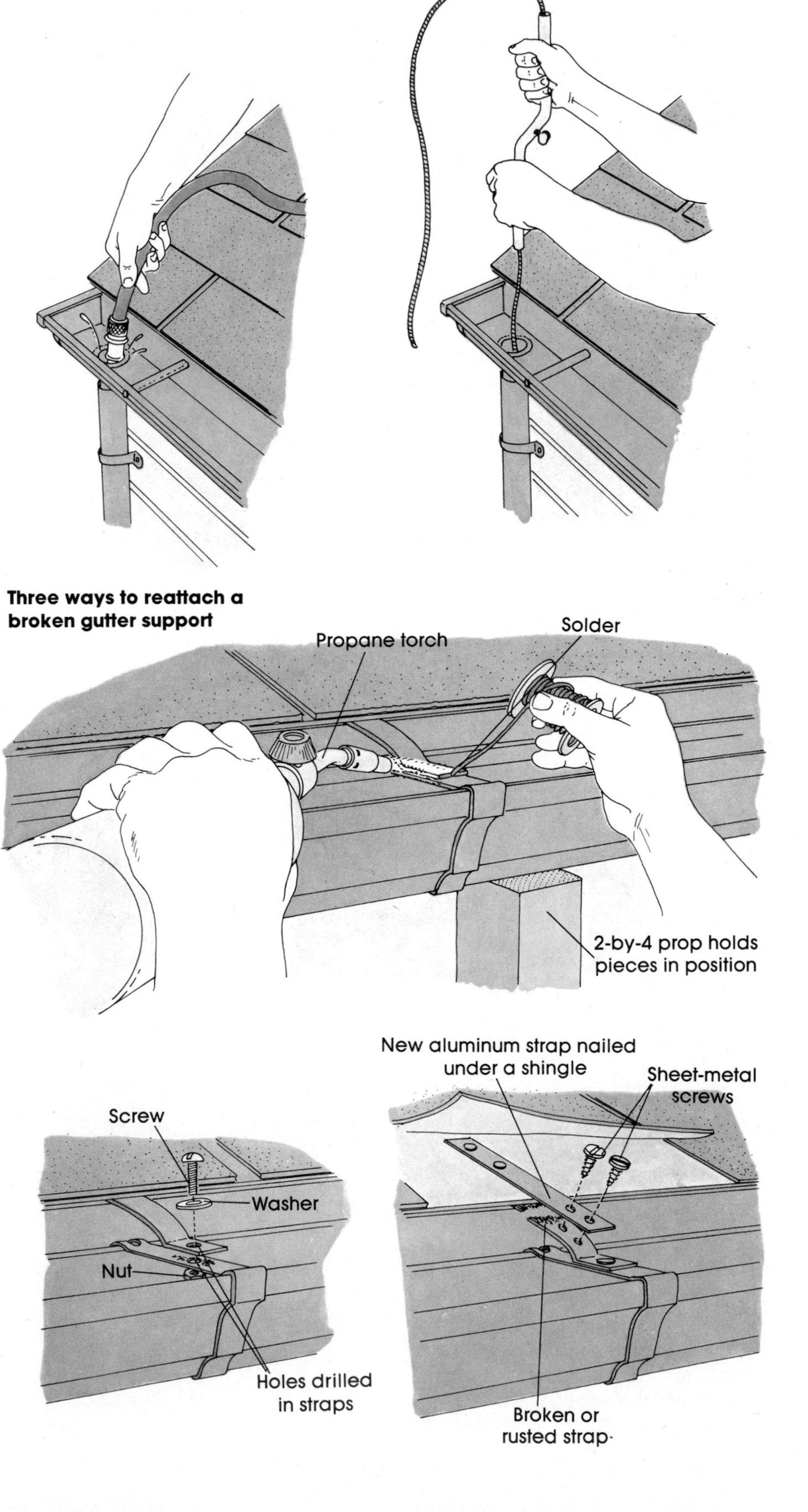

under it. Look at the type of gutter hanger you have and compare it to those illustrated on page 60.

☐ If the strap has separated from the hanger, drill a hole through the strap and hanger and join them with a nut and bolt.

☐ If the rivet holding the hanger to the gutter is broken, drill out the rivet and replace it with a galvanized nut and bolt.

☐ If the strap extending up under the roofing material is broken, cut it in half and remove the piece that extended under the roof. Replace it with a spike-and-ferrule hanger.

Holes in gutters. No matter what they are made of, gutters will eventually develop a hole through rust, rot, or puncture.

To repair holes in metal or vinyl gutters, first clean the area thoroughly with steel wool, working on the inside of the gutter. Small holes, the size of a nail, can be fixed with epoxy resin, available in most hardware stores.

Cover larger holes with adhesive-backed aluminum tape covered by a layer of roofing cement.

Wood gutters can develop holes through wood rot. To repair these, you must wait until the wood is thoroughly dry. Poke suspected areas with a screwdriver and watch for the soft, flaky wood that indicates rot.

Chisel out all the bad wood and then soak the hole with pentachlorophenol wood preservative. When it is dry, fill the hole with plastic wood and smooth carefully until it conforms to the gutter outline. Now give it a final protective coat of roofing cement.

Leaking joints. The final common problem with gutters is the joints. They separate because of loose hangers, ice loads, the weight of soggy leaves, and just old age. If you find a leaking joint, check for any loose hangers nearby and fix that problem first. On aluminum and galvanized steel gutters, use a caulking gun to squeeze silicone or butyl rubber caulking compound into the joint. With wood gutters, wait until the wood is completely dry and then caulk with butyl rubber compound.

Repairing leaky gutters

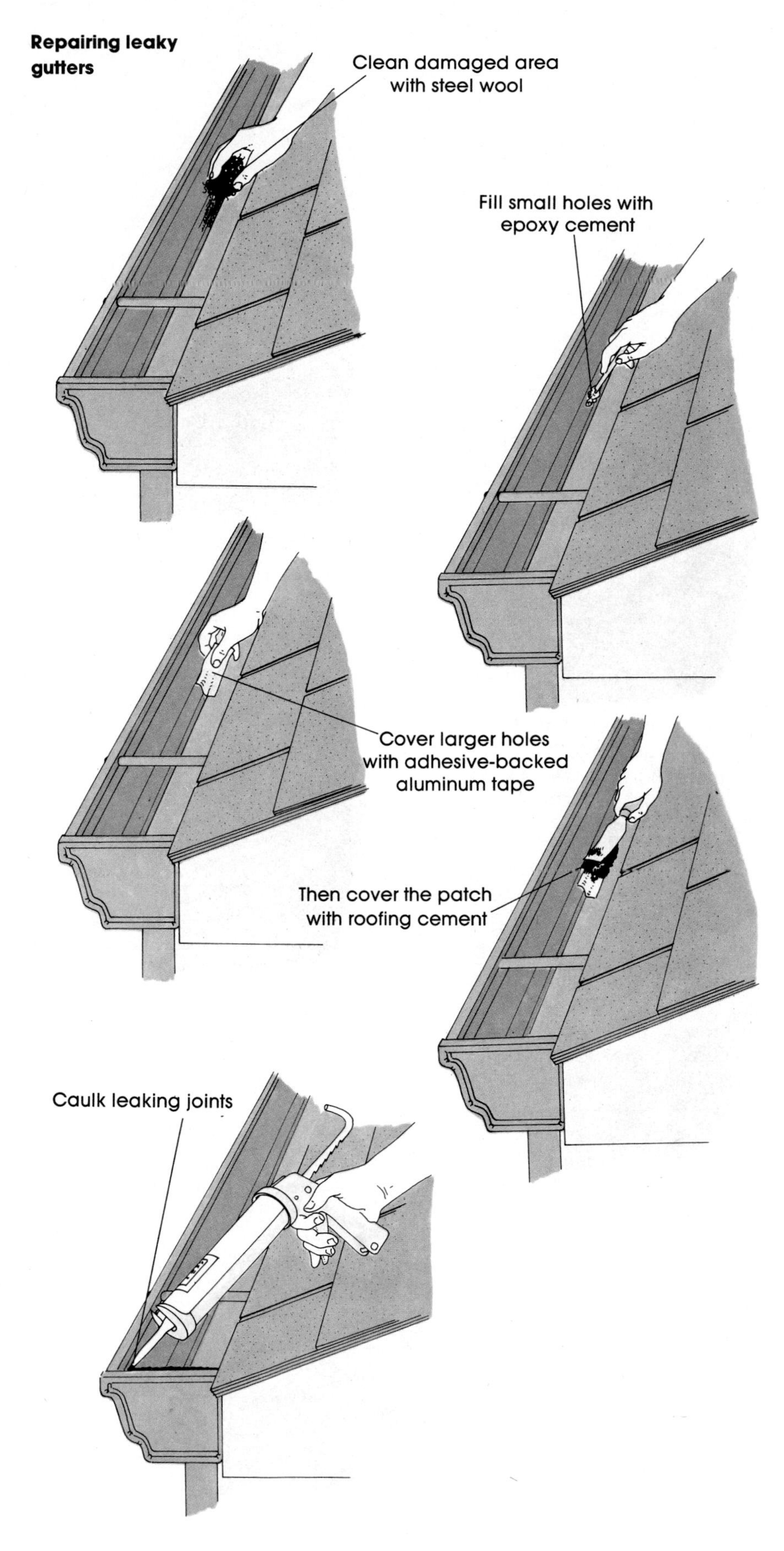

Board Siding

Repairing Split Boards

A siding board may split near an end where it was nailed. This should be repaired immediately to prevent rainwater from leaking into the stud cavity. Don't pull the nail where the wood split; you will mar the siding in the process. Use a chisel to pry the split further apart, then coat the interior of the split with an epoxy resin glue, which is waterproof.

When the glue is tacky, push the split together again. To prevent further splitting, use a push drill with a bit half the size of an 8d casing nail to drill holes on both sides of the split over the stud; nail. If there is no stud to nail to, try to drive shingle shims on both sides of the split to hold it until the glue dries. Set the nails; cover with wood putty that matches the house paint.

If you used a white glue, sand the glue surface lightly when it is dry and then repaint it, or the glue will not last through the winter.

Replacing Boards

If a siding board is badly splintered or rotted, it must be replaced.

Tongue and groove. To remove and replace a section of tongue-and-groove (T&G) boards, first locate the nearest studs on each side of the damaged area. This may be difficult since T&G siding is blind-nailed—that is, you can't see any exposed nail heads. One way is to use a stud finder, an inexpensive magnetic device that finds nails in studs. You can also measure from one corner of the house, keeping in mind that studs are generally placed every 16 inches on center. Tap the boards until you hear a solid rather than hollow sound.

Cuts in the board section to be removed must be made down the center of a stud so there will be backing for the new piece. Mark the line, then use a circular saw to make a pocket cut on the line. Do this by raising the blade guard, resting the saw on the front of the base plate, starting it, and gently lowering the blade into the board. Do not cut the adjacent boards; instead, finish the cuts with a chisel. Now split the section down the middle with the chisel, or cut it with the saw, and pry out. Remove any nails.

How to repair a split siding board

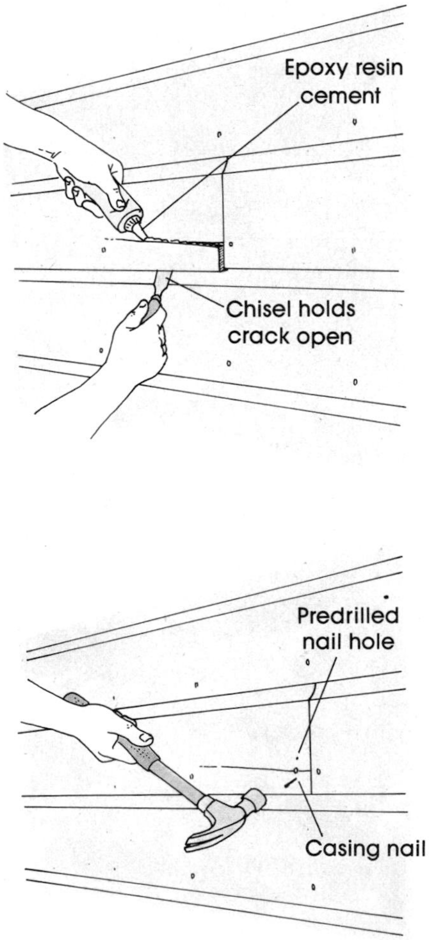

Epoxy resin cement

Chisel holds crack open

Predrilled nail hole

Casing nail

How to replace a piece of T&G siding

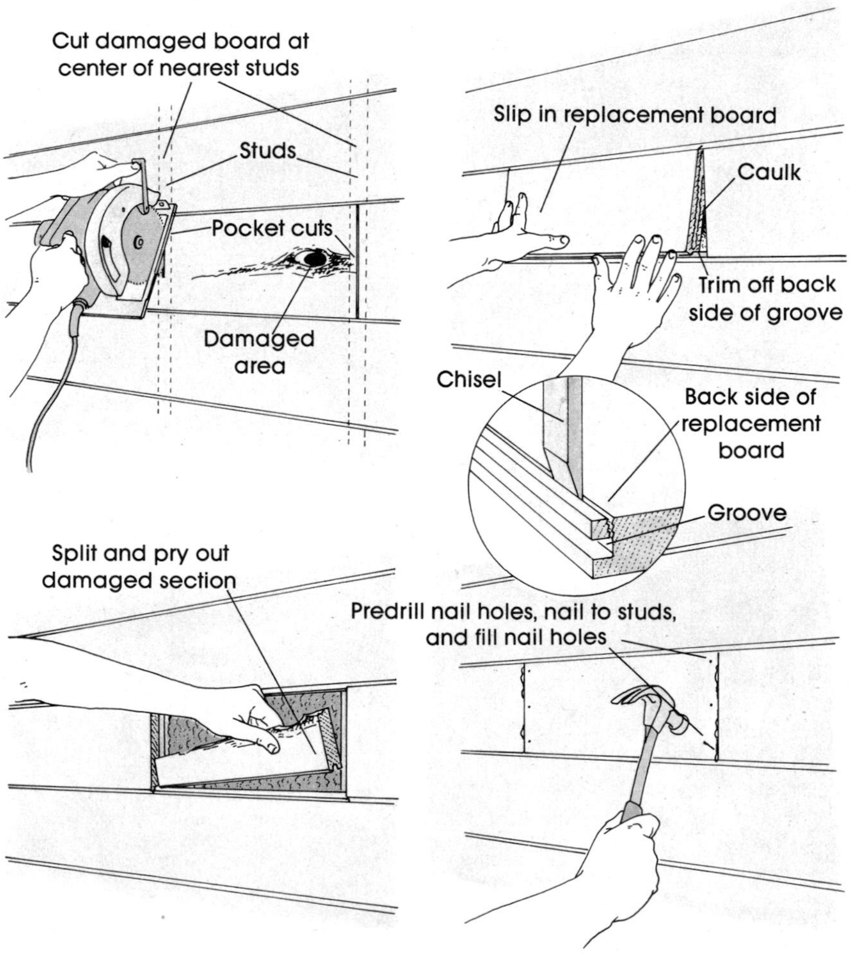

Cut damaged board at center of nearest studs

Studs

Pocket cuts

Damaged area

Split and pry out damaged section

Slip in replacement board

Caulk

Trim off back side of groove

Chisel

Back side of replacement board

Groove

Predrill nail holes, nail to studs, and fill nail holes

To slip the replacement board in place, first trim off the backside of the groove, as illustrated, then slip the board in place and nail it to the studs. Set the nail heads and cover with wood putty.

Lapped. Siding such as clapboard, shiplap, and other styles of lap siding is removed in much the same way as is tongue and groove. Cuts should be centered down the studs nearest each side of the damaged area. Lap siding is somewhat easier to remove because it is not locked in with a tongue-and-groove system. Use a chisel to pry the center part of the board out, then slip wedges of wood shingle under the siding near both ends to hold it out. Cut the boards with a circular saw as described above or with a back saw. Finish the cut with a keyhole saw. Caulk the edges of the new piece, then fit in place and nail.

Plywood Siding

Because of plywood's inherent strength, damage is not common. Should your siding be damaged, however, it is usually best to replace the entire panel, since a patched section will be quite noticeable. To replace a panel, remove any battens over the edges, pull the nails along the edge, and take off the panel. Nail the replacement in place, then paint or stain to match the rest of the siding.

Shake and Wood Shingle Siding

Repair damaged shake or wood shingle siding as you would a roof (see pages 86–87).

Stucco Siding

Cracks are the most common problem in stucco siding. To fix them, use the point of a can opener to gouge out the stucco under each edge of the crack in a bell shape. This provides a "key" for the repair material to lock into. Buy a small package of premixed stucco mortar and mix with water until it is a thick, creamy texture. Thoroughly wet the crack, then trowel mortar into it and smooth over. The mortar comes in colors, but you will probably have to paint that side for a finished appearance.

Metal or Vinyl Siding

Repairing metal and vinyl siding usually means removing a panel, which is probably best done by a professional installer.

How to replace a piece of lapped siding

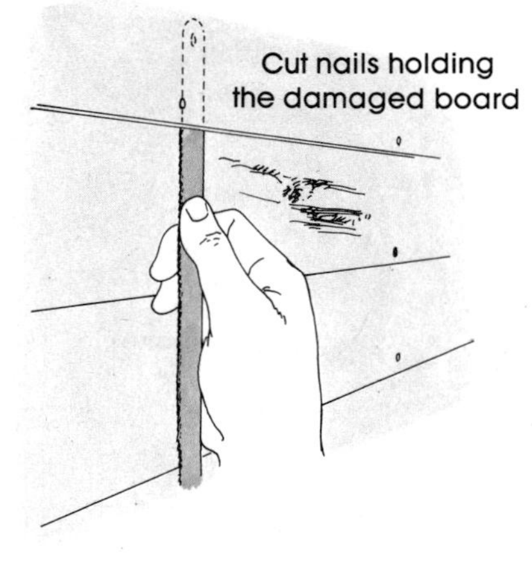

Wedge board out from others and saw out damaged part

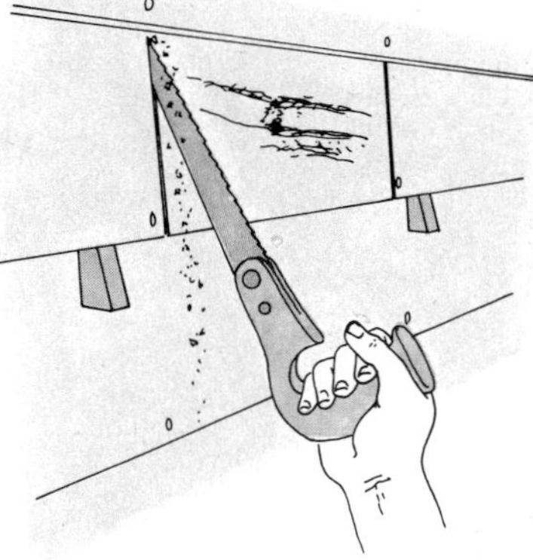

Predrill nail holes and nail in new board

Patching a crack in stucco

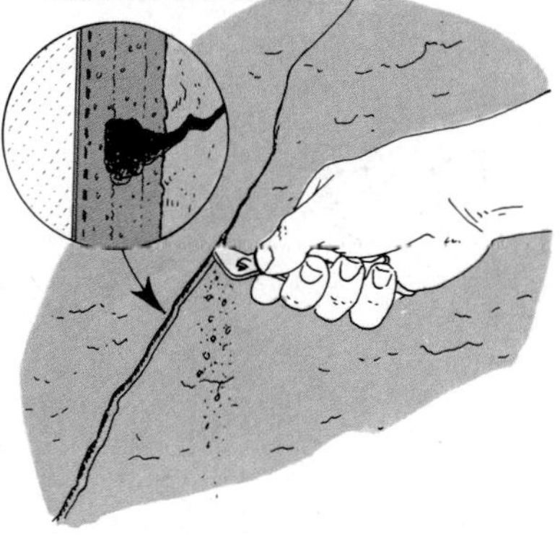

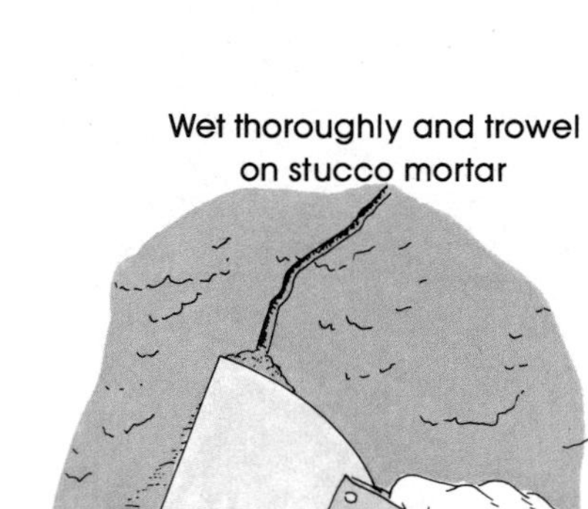

INDEX

Note: *Italicized* page numbers refer to illustrations.

A

Appearance of roofs, 24, 54
Asphalt Roofing Manufacturers Association, 25
Asphalt roll-roofing, 20
Asphalt shingle roof, 20, 22
 colors of, 25
 cost of, 24
 installing, 38–43
 over other roofs, 28
 patterns of, 38–40
 and pitch, 25
 removing, 29
 replacing, with wood shingles, 48
 re-roofing over, 28, 43
 sheathing for, 30
 step flashing on, 57
 vent flashing on, 55
Asphalt shingles: aligning, 36–37
 to check, 12
 to cut, 37, 40, 41
 fire resistance of, 24
 hip, cutting, 41
  estimating quantity of, 26
 loading a roof with, 32
 nailing, 36–37
 repairing, 86
 ridge, installing, 41
 and roof pitch, 25
 stacking, *32*
 warped, 28
 wind resistance of, 25
Attitude, and safety, 14

B

Beadboard. *See* Insulation, rigid
Blade, Carborundum, 44, 50, 54
Bonding agent, masonry, 80
Braces (wall), let-in, 68
 metal strap, 68
Brick, flashing for, 56
Bubble, in tar-and-gravel roof, 87
Building paper, 79
 for siding, 68

C

Caulking gun, *18*
Cedar, siding, 64, 69, 71, 72, 76
 shingles of, 47, 76
Ceilings, cathedral, 31
Cement, roofing, 26, 43, 51, 55
Chalkline, *18*
Charts: area/rake conversion, 26, 27
 color, 25
 hip and valley conversion, 26, 27
Chimney: inspecting, 85
 flashing around, 51, 57, 58–59
  to repair, 89
 installing tile around, 51
Chisel, *18*
 cape, 58
Clips, for tiles, 50
Clothing, suitable, 14
Codes, building, 13, 29
Color of roofing materials, 25
Composition shingles. *See* Asphalt shingles
Concrete. *See also* Mortar
 for tiles, 51
Contractor, roofing, working with, 17
Corner bead, 80
Corner trim: with horizontal board siding, 71
 with panel siding, 75
 with shingle siding, 77
 with vertical board siding, 73
Cost of roof, 13, 24
Coverage, of roofing materials, 22–23, 33
"Cricket" for chimney, 58–59
Crowbar, 29

D

Debris, dealing with, 29
Deck, roof. *See* Sheathing
Decoration, roofs and siding as, 5
Dolly Varden. *See* Siding, styles of
Doors, flashing for wood-framed, 68
Dormers: applying roofing felt over, 33
 calculating area of, 26, *27*
 flashing, 57
 shingling, 42–43
Downspouts: installing, 60–61
 maintaining, 85
Drip edge, 44, 47
 estimating quantity of, 26
 installing, 33, 38
 for tile roof, 49
Drywell, installing, 61
"Dubbing" shingle corners, 35
Durability, of roofs, 13

E

Eaves: drip edges for, 38
 inspecting, 12, 28
Exposure (weather) of shakes, shingles:
 on roofs: guide for, 47
  setting, 36
 on siding, 76

F

Feathering strips. *See* "Horsefeathers"
Felt, roofing, 22, 23. *See also* Underlayment
 applying, 33
 coverage of, 23, 33
 installing, for shake roof, 44
 under siding, 72
 for tile roof, 49
Fir siding, 72
Fire resistance of roofing materials, 24
Flashing: base, 58
 cap (counter), 58–59
 for chimney, 57, 58–59
 continuous, 56
 counter. *See* Flashing, cap
 estimating quantity, 26
 for hip roof, 40
 inspecting, 12, 85
 installing, 34–35, 38, 55–59
 repairing, 88–89
 roll-roofing, 34–35
 rubber-sleeve vent, *55*
 sheet-metal vent, *55*
 under siding, 68
 step, 57
 on tar and gravel roofs, 87
 two-piece plastic vent, *55*
 valley, installing, 34–35
 vent, 26, 46, 47, 51, 55
 W-metal, 34–35, 45, 49
 for wood-framed doors and windows, 68
Flashing paper for panel siding, 74
Float, wood, *78*, 82, 83
Furring strips, 70, 72

G

Gable ends, siding for, 71, 73, 75
Grades: of shakes/shingles, 22
Gutters: cleaning, 90
 estimating quantity of, 61
 inspecting, 85, 90
 installing, 61
 maintaining, 85
 repairing, 90–91
 types of, 60–61

H

Hammer, *78*
Hardboard siding, 64, 74
Hatchet, roofing (roofer's), *19, 36,* 37, 45
Hawk, *78*
Heating coils, for roofs, 89
Hips: applying shakes on, 46
 sheathing on, 31
 shingling, 40–41, 48
 roll-roofing on, 53
 installing tile on, 49, 50–51
"Horsefeathers" (feathering strips), 28

I

Inspecting, for maintenance, 12, 85
Insulation: checking, 12
 of panel roofs, 23, 54
 and re-siding, 63, 69
 rigid (beadboard), 63
 and tile roofs, 49
 under vertical siding, 72
Insurance, 15, 17

J

Jamb extenders, 66–67

K

Kraft paper, 76

L

Ladder jacks, *15*
Ladder loader, 32
Ladders, suitable types of, 14–15
Lath sheets, 80
Leader. *See* Splash block
Leak, to find, 12–13, 86
Level, *78*
Lumber, "green," 72, 81

M

Maintenance, importance of, 85. *See also* Repairs
Materials for roofing, 20. *See also individual types*
 to estimate quantity of, 26–27
Metal: flashing. *See* Flashing
 panels, roofing. *See* Panels, metal
 siding. *See* Siding
Mitering, of corners on shingle siding, 77
Mortar, Portland cement, 58
 mixing, 59
 repairing, 89
Mortar board, 81
Mortar box, *78*, 81

N

Nails: aluminum, 38
 furring, *80*
 estimating quantity of, 26
 for metal flashing, 56
 for panel roofs, 54
 for panel siding, 74
 for shingling, 38, 43
  wood, 47
 for tile roofs, 50
 for wood siding, 71
Nailing: asphalt shingles, 36
 concealed method for roll-roofing, 53
 exposed method for roll-roofing, 52–53
 patterns for wood siding, *71, 72*
 shingle siding, 76
 wood shingles and shakes, 36
Nail stripper, 37

P

Paint: damage to, 63
 exterior, to check, 12
 and siding, 76
Panel roof: appearance of, 54
 installing, 54
 repairing, 87
 vent flashing on, 55
Panels, roof: 23, 54
 cost of, 24
 cutting, 54
 fire resistance of, 24
 installing, 54
 painting, 54
 and pitch, 25
 re-roofing with, 28
 sheathing for, 30–31
 vent flashing for, 55
 wind resistance of, 25
Panels, siding, *64*, 74–75
 appearance, 63
 cost, 63
 cutting, 75
 installing, 74–75
 repairing, 93
 styles, 64
 wall preparation for, 68
Pattern: for asphalt shingles on roof, 25, 38–40
 following original in re-roofing, 43
 for wood shingle siding, *65*
Permit, building, 13, 29
Pine siding, 72
Pitch (slope, steepness) of roof: to determine, 21
 extreme, to shingle, 43
 and roofing materials, 25
 and safety, 15
 and shakes, 44
 and tiles, 49, 50
 and wood shingles, 47
Plastic sheeting, 26
Plywood: sheathing for roof, installing, 30–31
 sheathing for wall, 79
 siding, 64, 74–75. *See also* Panels, siding
Pocket cut, to make, *75*
Pry bar, *18*
Putty knife, *18*

Q

Quality of roof, 24

R

Rafters: inspecting, 12
strength of, and roofing materials, 25
Rakes: drip edges for, 38
Rake trim for panel roof, 54
Red resin paper, 76
Redwood siding, 64, 69, 71, 72
Removing an old roof, 29
Repairs. *See also* Inspection; Maintenance
to gutters and downspouts, 90–91
to roofs, 86–89
to siding, 92–93
Re-roofing, 28–29
with asphalt shingles, 43
with wood shingles, 48
Re-siding, 63
with panels, 74
preparation for, 69
Re-stuccoing, preparation for, 80
Ridge cap, for panel roofs, 54
Ridges: asphalt shingles on, 41
roll-roofing on, 53
roofing felt on, 33
shakes on, 46
sheathing on, 31
tile on, 49, 51
wood shingles on, 48
Roll-roofing, 22–23
cost of, 24
coverage of, 23, 52
fire resistance of, 24
to install, 52–53
loading on roof, 32
open valley, 34–35
over old roofs, 28
and pitch, 25
repairing, 87
re-roofing over, 48
sheathing for, 30
vent flashing on, 55
wind resistance of, 25
Roof. *See also individual types*
built-up. *See* Tar and gravel roof
calculating area of, 26–27
cold-mopped, 24
flat, to inspect, 12
hot-mopped, 23
inspecting, 12, 28, 85
loading, 32
mansard, shingling, 43
on new construction, 30–31
planning for, 13
as protection, 11
repairing, 86–89
replacing, 28–29
removing, 29
shed, 53, 54
styles of, 20
Roof deck. *See also* Sheathing
to inspect, 28
Roofer's seat, to make, 45
Roof jacks, using, 19
Roofing felt. *See* Felt, roofing
Rubber sleeve flashing, *55*

S

Safety, guidelines for, 14–15
while shingling, 36
Saw: circular, 37, 50, 54
hand, *18*
saber, 37
Scaffolding, *14, 15*
Scarifier, *78,* 81
Sealant, asphalt, 58
Shake roof, 20
continuous flashing with, 56
inspecting, 12
installing, 44–46
pitch of, 25
removing, 29
sheathing for, 30, 31
step flashing for, 57
vent flashing for, 55
Shakes, 22
aligning, 37
applying, 45–46
coverage of, 22, 44
cutting, 37
fire resistance of, 24
loading on roof, 32
nailing, 36
re-roofing with, 28
repairing, 86
types of, 44
wind resistance of, 25
Sheathing (roof deck), 30–31
to inspect, 12
materials for, 30–31
for roll-roofing, 53
for shake roof, 44
solid, 30–31
spaced, 31, 44
for tile roof, 49
for wood shingle roof, 47
Sheathing (wall): for shingle siding, 76
for stucco, 79
Shingle roof. *See* Asphalt shingle roof; Wood shingles
Shingle siding, 64–65
estimating quantity of, 76
fancy-cut, *65,* 76
"green," 76
installing, 76–77
to trim, 77
Shingles: aluminum, 24
asphalt. *See* Asphalt shingles
composition. *See* Asphalt shingles
wood. *See* Wood shingles
Shingling, techniques of, 36–37
Shovel, flat-bottom, 29, *78*
Siding: aluminum, 63, 65
bevel, *64*
board, 63
to install, 69–73
to repair, 92–93
styles of, *64*
board-and-batten, *64,* 72
brick, 63
bungalow, *64*
choosing, 63–65, 69, 73
colors of, 25
drop, *64*
hardboard, 64
to install, 74–75
inspecting, 12, 85
maintenance of, 63, 85
metal, 65
to repair, 93
new over old, 63, 66–67, 69, 72, 80
panel, 63, 64
to install, 74–75
planning for, 13
plywood, 63, 64
to install, 74–75
to repair, 93
styles of, *64*
preparation for, 66–68, 69–70, 72, 76, 79–80
as protection, 11
rabbet edge (Dolly Varden), *64*
to repair, 92–93
shingles for, 22, 63, 64
to install, 76–77
to repair, 93
styles of, *65*
shiplap, *64*
steel, 63, 65
styles of, 64–65
tongue and groove. *See* Tongue and groove boards
vinyl, 63, 65
to repair, 93
wood, 64–65
Skylights, calculating area of, 26
Slate, for roofs, 24
cost of, 24
fire resistance of, 24
removing, 29
replacing, 28
sheathing for, 30
wind resistance of, 25
Slope. *See* Pitch
Specialist, roofing, working with, 17
Splash block (leader), *60,* 61
Squangle, 21
Staple (nailing) gun, 37
Starter board: for horizontal siding, 70
for shingle siding, 76
Starter board (V-rustic), for roof sheathing: installing, 30
Starter course: shakes, 45
wood shingles, 47
Starter roll for asphalt shingles, 38
estimating quantity of, 26
Starter strip: asphalt shingles as, 38
estimating quantity of, 26
for tile roof, 49–50
Steepness of roof. *See* Pitch
Step flashing, to make, 57
Stop bead, 79
Story pole: to make, 67
to use, 70, 76
Strainer basket for gutter, *60,* 61
Stucco, 63
applying, 78–83
technique for, 81
brown coat, 78, 81–82
finish coat, 78, 83
flashing with, 56
to mix, 80–81
Modern American finish, *65,* 83
Old English finish, *65,* 83
finish patterns for, *65,* 83
preparation for, 79–80
to repair, 93
scratch coat, 78, 81
screeds, applying, 82
spatter (spatter-dash) finish, *65,* 83
standard smooth finish, 83
tools for, *78*
travertine finish, *65,* 83
Stucco mixer, *78*

T

Tape measure, *18*
T&G. *See* Tongue and groove boards
Tar and gravel (built-up) roof, 20, 23
cost of, 24
fire resistance of, 24
inspecting, 12
repairing, 86–87
re-roofing, 28
with asphalt shingles, 43
with wood shingles, 48
wind resistance of, 25
Temperature, and installing roll-roofing, 52
Terminology, *10*
Tile roof: cost of, 24
installing, 49–51
nailing, 50
and pitch, 25
repairing, 87
replacing, 28
removing, 29
sheathing for, 30, 31
step flashing with, 57
vent flashing on, 55
weight of, 49
Tiles: clay, *23*
color of, 49
concrete, *23,* 49
installing, 49–51
cutting, 50
fire resistance of, 24
pattern of, 49
re-roofing with, 28
Spanish, *23,* 49
stacking, *32*
wind resistance of, 25
Time, estimating, 13, 17
Tin snips, *18, 78*
Tongue and groove boards (T&G): for roof sheathing, 31, 49
for siding: to install, 69–73
patterns of, *64*
to replace, 92–93
Tool belt, *18*
Tools, roofing, 18–19, *36, 37*
Trim. *See also* Corner trim
inspecting, 12
Trimming shingle corners, 35
Trowel, steel, *78*
"Tying-in" shingles above dormer, 42–43

U

UL. *See* Underwriters Laboratories
Underlayment, 22, 23. *See also* Felt, roofing
Underwriters Laboratories (UL), 24, 25
Utility knife, *19,* 37

V

Valley/valleys: applying roofing felt over, 33
to calculate materials for, 26, *27*
flashing, 34–35
on shake roofs, 45
on tile roofs, 49
full-lace, 35
half-lace, 35
metal open, 34
inspecting, 12
repairing, 88
roll-roofing open, 34
roll-roofing on, 52, 53
tile on, 50–51
wood shingles on, 48
Vapor holes, 69
Vents: applying roofing felt around, 33
applying shakes or wood shingles around, 46–47
flashing, 55
repairing flashing around, 88–89
applying tile around, 51
Vinyl siding. *See* Siding, vinyl
V-rustic. *See* Starter board

W

Wall: bracing, 68
continuous flashing on, 56
preparation, 66–68
for horizontal wood siding, 69–70
for vertical wood siding, 72
for shingle siding, 76
for stucco, 79–80
Weather. *See* Exposure
Wheelbarrow, *78*
Wind resistance, of roofing materials, 25
Windows: flashing for wood-framed, 68
jamb extenders for, 66–67
Wood shingles: aligning, 37
continuous flashing with, 56
cost, 24
coverage of, 22
fire resistance of, 24
installing, 47–48
nailing, 36
nails for, 47
and pitch, 25
removing, 29
repairing, 86
re-roofing over, 28, 43, 48
re-roofing with, 28, 48
sheathing for, 30–31
step flashing with, 57
vent flashing with, 55
wind resistance of, 25

METRIC CHART

U.S. Measure and Metric Measure Conversion Chart

	Formulas for Exact Measure				Rounded Measures for Quick Reference				
	Symbol	When you know:	Multiply by:	To find:					
Mass (Weight)	oz	ounces	28.35	grams	1 oz			=	30 g
	lb	pounds	0.45	kilograms	4 oz			=	115 g
	g	grams	0.035	ounces	8 oz			=	225 g
	kg	kilograms	2.2	pounds	16 oz	=	1 lb	=	450 g
					32 oz	=	2 lb	=	900 g
					36 oz	=	2-1/4 lb	=	1000 g (1 kg)
Volume	tsp	teaspoons	5	milliliters	1/4 tsp	=	1/24 oz	=	1 ml
	tbsp	tablespoons	15	milliliters	1/2 tsp	=	1/12 oz	=	2 ml
	fl oz	fluid ounces	29.57	milliliters	1 tsp	=	1/6 oz	=	5 ml
	c	cups	0.24	liters	1 tbsp	=	1/2 oz	=	15 ml
	pt	pints	0.47	liters	1 c	=	8 oz	=	250 ml
	qt	quarts	0.95	liters	2 c (1 pt)	=	16 oz	=	500 ml
	gal	gallons	3.785	liters	4 c (1 qt)	=	32 oz	=	1 l
	ml	milliliters	0.034	fluid ounces	4 qt (1 gal)	=	128 oz	=	3-3/4 l
Length	in.	inches	2.54	centimeters			3/8 in.	=	1 cm
	ft	feet	30.48	centimeters			1 in.	=	2.5 cm
	yd	yards	0.9144	meters			2 in.	=	5 cm
	mi	miles	1.609	kilometers			12 in. (1 ft)	=	30 cm
	km	kilometers	0.621	miles			1 yd	=	90 cm
	m	meters	1.094	yards			100 ft	=	30 m
	cm	centimeters	0.39	inches			1 mi	=	1.6 km
Temperature	F°	Fahrenheit	5/9 (after subtracting 32)	Celsius			32°F	=	0°C
							68°F	=	20°C
	C°	Celsius	9/5 +32	Fahrenheit			212°F	=	100°C
Area	in.^{2}	square inches	6.452	square centimeters			1 in.^{2}	=	6.5 cm^2
							1 ft^2	=	930 cm^2
	ft^2	square feet	929	square centimeters			1 yd^2	=	8360 cm^2
							1 a	=	4050 m^2
	yd^2	square yards	8361	square centimeters					
	a	acres	.4047	hectares					